Ionic Polymer Metallic Composite Transducers for Biomedical Robotics Applications

Andrew J. McDaid and Kean C. Aw

Ionic Polymer Metallic Composite Transducers for Biomedical Robotics Applications

International Frequency Sensor Association Publishing, S. L.

Andrew J. McDaid and Kean C. Aw
Ionic Polymer Metallic Composite Transducers for Biomedical Robotics
Applications

ISBN-13: 978-84-616-7669-9
ISBN-10: 84-616-7669-6
BN-20140130-XX
BIC: MQW

Acknowledgements

I would like to extend my sincere thanks to Prof. Kean Aw, who has been an exceptional mentor to me throughout my PhD research and to this day. His enthusiasm for research has been contagious and I greatly appreciate all the time we spend bouncing ideas during intellectual (and other) discussions. His support and guidance has been incredible.

I would like to thank all my colleagues at The University of Auckland who have been there for me whenever I have sought guidance or assistance. Although there are too many to name here, I would however like to make special mentions of Prof. Shane Xie and Prof. Enrico Haemmerle, who have been there throughout my research and studies; their expertise has been much appreciated. To Logan Stuart, for our many discussions and his readiness to give his advice and assistance at any time. Also to my former colleagues, Sean Manley, David Liu, Kaval Patel and Wei Yu, who I worked alongside with to carry out this research.

I would like to acknowledge The University of Auckland for the financial support they have provided me with both The University of Auckland Doctoral Scholarship as well as The University of Auckland Undergraduate Scholarship which I have been awarded.

Last, but far from least, I would like to thank my big family and all my friends for their continued support and encouragement. I am especially grateful for all the assertive motivation my wife Sarah constantly provides me; this has directed me to produce a piece of research work I am truly proud of, in a timely fashion! To my parents, thank you for instilling in me a desire to learn and excel in everything I do. And for everything else ...

Dr. Andrew McDaid,

Auckland, NZ

About the Authors

Andrew J. Mcdaid is a Lecturer (Assistant Professor) in the Department of Mechanical Engineering at The University of Auckland, New Zealand. His main research interests include intelligent mechatronics systems and devices, especially in rehabilitation, bio-mechatronics, bio-medical and surgical robotics applications. Andrew has application experience in 'smart' materials as well as biomedical, rehabilitation and traditional robotics fields. He is working to develop medical devices, using non-traditional sensors and actuators, which are bio-cooperative and work in harmony with the human brain and physical body.

Kean C. Aw is an Associate Professor in the Department of Mechanical Engineering at The University of Auckland, New Zealand. His main research interests include micro-systems and deployment of smart/functional materials such as conducting polymers, metallic oxides, etc as sensors and actuators in various applications such as bio-sensors, medical/rehabilitation robots, micro-pumps, micro-manipulators, MEMS, energy harvester, etc.

Contents

Preface

Robotic devices have traditionally been developed for industrial applications for tasks which are repetitive, inhospitable and even unachievable by humans. The natural progression then for future robotic devices is to be intelligent so they can work closely with humans in their own environment. This book is written for leading edge engineers and researchers, working with non-traditional or smart material based actuators, to help them develop such real world biomedical applications. Electrical, mechanical, mechatronics and control systems engineers will all benefit from the different techniques described in this book. The book may also serve as a reference for advanced research focused undergraduate and postgraduate students.

Specifically, this book describes a cluster of research which aims to not only advance the state of art through scientific progress in a specific smart material actuator, namely IPMC, but also serve as a guideline to demonstrate the techniques in which many more issues around developing future smart material actuators can be solved. Traditionally actuators are well known and understood and so designing mechanical devices is almost trivial, however developing 'smart' devices for complex medical applications requires designing from a fundamental standpoint. This research-design-development process is described in this book.

To this end, six biomedical device prototypes have been developed, by first creating a new physics based, design oriented model of the IPMC actuators themselves, in order to be able to completely simulate the system and prove the design before committing to implementation. Following from this, new controller algorithms (specific for each application) are developed, which use the fundamental IPMC model coupled with the mechanism dynamics model, in order to control the extremely complex, nonlinear and time-varying IPMC system.

Overall, IPMCs (and typically all smart materials) have many advantages over traditional actuators and in my opinion are the key to advancing from the current state of the art to a new level of bio-compatible systems which work in harmony with the human body. However, much in-depth research, both fundamental and applied, is needed into the development of such systems to make them viable. Simply selecting an 'off the shelf' actuator and designing a system around the chosen actuator, as is done in most current system design, will simply not work. The framework in this book outlines a method of system design, with smart material actuators, in order to develop smart devices which have the potential to improve the health of society in the future, which is what I am really passionate about.

Book Synopsis

To start, the background of this research is introduced and the great need for new actuator technologies is discussed. Fundamentals of IPMCs and biomedical robotics are also presented, followed by the objectives and scope of the research published in this book.

A thorough survey of the state of art in IPMCs is reported in Chapter 2. This review uncovers the areas of IPMC research which are lacking and hence require further investigation to further knowledge in this field. After this review the contributions of the research in this book are fully justified.

Chapter 3 describes the development of a new comprehensive scalable electro-mechanical IPMC model which has been created specifically for designing mechanical systems. This is used as an invaluable tool for the remainder of the research as it is used to evaluate the performance of all the devices before they are implemented as well as for determining a suitable stable starting state for the controller tuning algorithms.

A bio-inspired rotary device which replicates the performance of a traditional stepper motor is presented in Chapter 4. First the motor design with open-loop control is simulated using the IPMC model to verify its performance and then experimental validation is presented on the real system. An extension to the motor mechanical design to improve performance is also proposed.

The fundamentals and state of the art in iterative feedback tuning is presented in Chapter 5 along with experiments demonstrating the capabilities of the existing iterative feedback tuning algorithm to tune IPMC performance. The iterative feedback tuning approach is the basis for the new closed-loop control system architectures developed and used for the remaining applications.

Chapter 6 presents the design and implementation of an artificial muscle actuator driving a rotary joint which is envisaged for use as a prosthetic finger joint or a hand rehabilitation device. As the human finger joint requires large deflections, the IPMC becomes highly

17

nonlinear and as such a gain scheduled (GS) controller tuned with iterative feedback tuning is developed. Experimental results show its superior performance to a linear proportional-integral-derivative (PID) controller.

In Chapter 7 an IPMC actuated micropump for dispensing drugs to humans is presented. The micropump has the ability to be embedded into a human. A modification to the iterative feedback tuning algorithm which enables the system to be tuned online is developed to adaptively tune the micropump throughout its operation.

Chapter 8 describes a microtool/gripper and micromanipulator. The devices are first designed to meet their required specifications through a thorough design process utilizing the new IPMC model and dynamic mathematical models of the mechanical systems to simulate their performances. New robust controllers, which are adaptively tuned using a modified iterative feedback tuning algorithm, are developed for the devices. The devices are then implemented and experimental results of the entire system to verify its performance are presented.

Chapter 9 demonstrates the use of IPMC as a surgical robot end effector. The use of IPMC that is compliant can add safety to delicate surgery. A simple force compliant surgical robotic tool has been implemented and the experiments conducted show promising results for using IPMCs in real world applications.

The conclusions and summary of this work are presented in the final chapter.

Chapter 1

Introduction

Robotic devices have traditionally been developed for industrial applications to enhance productivity, reliability and accuracy for tasks which are repetitive, inhospitable and even unachievable by humans. Over time the capabilities and hence complexity of these devices has been exponentially increasing, mainly due to developments in computing speed and capacity which has led to substantial advances in control theory, machine learning and artificial intelligence.

For these reasons many of the problems for industrial robotics have been solved and this is now a well understood field. The natural progression then for future robotic devices is to work closely with humans in their own environment and even become embedded with humans themselves for improving quality of life. Some such examples of existing biomedical devices include hearing aids, pacemakers and prosthetic limbs.

1.1. The Need for New Actuator Technologies

Traditional robotic actuators, such as thermo-chemical motors (combustion engines), electromagnetic drives, and hydraulic/pneumatic machines, have all been extensively investigated. Although these devices and their control systems are well understood and have advanced performance which can, in some aspects, surpass that of humans, they simply do not have the capabilities and diversity required to meet the demand for new actuators in a large variety of new applications in areas such as mechatronics and biomedical robotics [1-4].

The major restriction now to the progression of future biomedical robotic devices is suitable mechanical transducers.

Biological muscles have gained much attention as actuators due to their functional properties and many regard these muscles as an ideal example of an actuation system. Conventional actuators, however, differ majorly from natural muscle not only in weight, rigidity, and compliance but even in their actuation principles. Attempts have been made through implementing active compliance algorithms and coupling actuators with passive elements (mechanical springs etc.) to provide safety but this is simply masking the fact that none of these traditional components are suitable for mimicking biological systems or interacting flawlessly in a human environment.

To illustrate the need for new approaches to actuator technologies consider the state of the art in human arm and leg prostheses which are driven by electric motors. Despite years of intensive research they are still stiff, heavy, noisy and extremely clumsy with far to go before true integration with humans is possible. These devices would benefit a great deal from new actuators which have similar properties to biological muscles.

Another consideration is the demand to manipulate smaller objects and miniaturize devices, for example insect inspired flying and crawling robots. Traditional actuators, due to fabrication limitations and driving principles as well as their power-to-weight ratios, have limits preventing them from being scaled down to the sizes required to build such devices. Even through miniaturization and on-going incremental research, these traditional transducers will not replicate biological systems [2].

To truly operate seamlessly with humans and mimic biological systems, robotic devices must possess similar actuation characteristics to real biological systems.

To achieve this it is clear that a totally new approach must be taken when designing biomedical robotic devices. Bio-inspired transducers which have similar properties to human tissue and muscle, in particular mechanical compliance, structural simplicity to allow easy scalability, high power-to-weight and power-to-volume ratios, precise and embedded control capabilities, must be developed. New electro-active-polymer (EAP) materials are demonstrating promise in this area and are envisaged by many researchers as the way forward for developing robotic systems which can truly mimic human characteristics.

Some electronic EAPs, for example dielectric based elastomer actuators, have demonstrated high forces and properties nearing human capabilities. The main drawback with this class of EAP is the very high voltages required (>1,000 V), this makes their integration with humans questionable. Ionic polymer metallic composites (IPMCs) are a type of ionic EAP which have extensive desirable characteristics when compared with traditional actuators and whose actuation mechanisms can mimic biological muscle. With the advancement of materials science research, as well as actuator modeling and control, these IPMC transducers are working towards human capabilities.

1.2.   IPMCs: Fundamentals

IPMCs are a type of smart material transducer. They are a class of EAP which act as an actuator under the influence of an electric field and conversely produce an electric potential when mechanically deformed. Typically IPMCs are fabricated in strips and are operated in a cantilever configuration where a voltage is either applied or measured at the base through a set of clamped electrodes, see Fig. 1.1. A beam type actuation greater than 90° can be achieved with small applied voltages, typically less than 5 V. The sensing voltage is usually 1-2 orders of magnitude lower than voltages required for actuation [5].

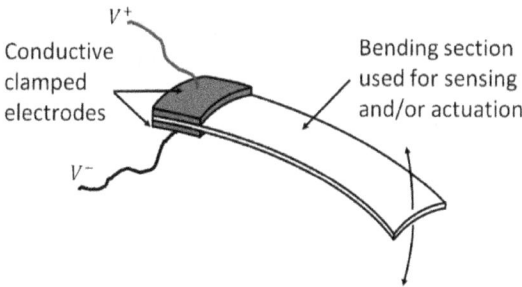

Fig. 1.1. Schematic drawing of an IPMC transducer
in cantilever configuration.

IPMCs are fabricated with a perfluorinated ionic membrane, for example Nafion® by DuPont, which is sandwiched between two thinly coated conducting electrodes of a noble metal, typically platinum or gold, on either side of the polymer, see Fig. 1.2. The ionic polymer must be an ion-exchange membrane that is permeable to cations but not anions, i.e. consisting of a fixed network of anions with mobile cations. The IPMC must be hydrated before and, depending on the polymer matrix and electrodes, sometimes during operation.

21

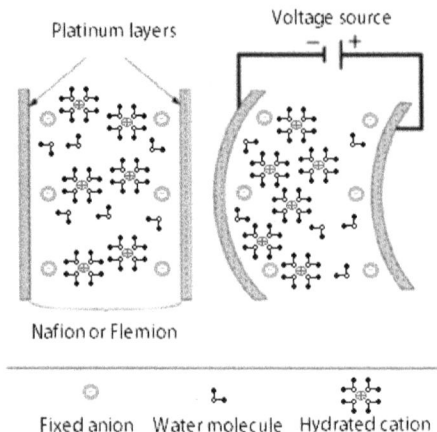

Fig. 1.2. Cross-section of an IPMC transducer (modified from [6]).

When an electric field is applied to the clamped electrodes at the base, a voltage is induced along the thinly coated metal electrodes and hence across the entire length of the polymer. As a result of this electrical potential the hydrated mobile cations migrate to the cathode. The accumulation of water on one side of the polymer causes expansion which in turn produces a differential stress across the thickness; the uneven stress then produces a mechanical strain or bending deformation. The opposite effect results in the sensing phenomena of an IPMC. The main transduction mechanism is summarized in Fig. 1.3.

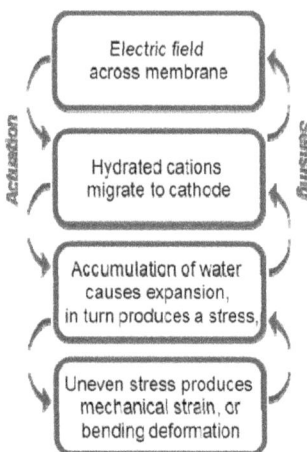

Fig. 1.3. Summary of the transduction mechanism for IPMCs.

When actuated with a constant DC electric field, the actuator will eventually relax back towards the origin as the loose water diffuses back, this is known as 'back-relaxation' [5, 7, 8]. Due to the need for hydrated ions, IPMC have traditionally operated best in aqueous environments although new ionic solvents have been investigated which enable proficient operation in air [9 - 11]. Some more recently developed IPMCs operate better in air as the loose water is evaporated leaving only the hydrated ions, this has been shown to alleviate the back relaxation phenomena.

IPMCs are most commonly manufactured in sheets where individual transducers are cut from the sheet; as such the possible geometries are infinite and can be tailored to any application requirements [12, 13]. This makes IPMCs highly suited for implementation into any number of biomedical and mechatronics systems as well as making them easily scalable for miniaturization applications as well as in MEMS type devices. IPMC performance characteristics are highly dependent on their geometry [14 - 17]; this highlights the importance of developing accurate scalable models for designing applications with IPMCs. IPMC materials have a number of superior properties, in comparison with traditional actuators and other EAPs, which makes them highly desirable for use as mechanical actuators, including:

i.   Lightweight and thin, typically mass less than 1 g and between 200 μm to 2 mm thick.

ii. Flexible and compliant, hence safe for operating in sensitive biological environments.

iii. Biocompatible and implantable in humans [15].

iv. Low actuation voltages make them safe for operating near humans, typically ± 1 to 5 V.

v.   High mechanical displacement (>90°) and force-weight ratio (force > 60 × weight).

vi. Can operate at high speeds, up to 100 Hz [18] (bandwidth is highly dependent on thickness of IPMC).

vii. Mechanically simple due to the direct conversion of electrical to mechanical energy.

viii. Easy scalability without the loss of actuation and sensing capabilities.

ix. Can achieve both micro and macro deflections without any gearing mechanisms [19].

x. Integration of sensing and actuating using the same device [20 - 22].

xi. Low power consumption, hence good for embedded and remote applications.

xii. Fully operational underwater (tested up to the equivalent of the deepest ocean on earth, 10,000 m [23]), at low temperatures and in vacuum [24, 25].

xiii. Fabricated to many shapes, for example helical [13], disc type [26], finger shape [27], rectangular tubes [28].

xiv. Durable and chemically stable, possible to bend over 106 times [18].

xv. Completely noiseless actuation, unlike electric motors or pneumatics.

Table 1.1 outlines some of the IPMC properties in comparison with some key characteristics of MSM as a reference. MSM is an incredibly elegant system that is a challenge to emulate. Muscle is a 3D nanofabricated system with integrated sensors, energy delivery, waste/heat removal, local energy supply, and repair mechanisms. IPMCs are still in their infancy and so some of their properties are relatively far from MSM. Although with continued materials science research they have the potential to display equivalent properties to MSM, unlike many traditional actuators. The data in Table 1.1 is based on literature in [29 - 32]; some data is not readily available and has therefore been omitted from this table.

The cost of IPMCs varies and depends mainly on the electrode material and thickness as the electrodes are typically fabricated from precious metals such as gold or platinum. This can make IPMCs reasonably expensive due to the increasing price of such materials. In 2005 it was reported in [33] the cost of IPMCs purchased commercially was US $ 100,000/kg. Since then the cost has increased. At the time of writing this book the average purchase price of a typical IPMC actuator 30 mm long by 10 mm wide is between US $ 200 to 250 [33].

Despite the many advantages of IPMCs there are still a number of performance issues that need to be overcome before IPMCs are widely regarded as viable alternatives for traditional actuators in real life biomedical and industrial applications.

Table 1.1. IPMC properties in comparison with human muscles.

Property	Typical		Maximum	
	MSM	IPMC	MSM	IPMC
Strain (%)	20	0.5	< 40	3.3
Stress (MPa)	0.1	3	0.35	15
Work Density (kJ/m^3)	8		40	5.5
Strain Rate (%/s)			> 50	3.3
Specific Power (W/kg)	50		284	2.56
Efficiency (%)		1.5	40	2.9
Modulus (MPa)	10-60	100		
Density (kg/m^3)	1,037	1,500		
Life Cycle			> 10^9	10^6

Some of the major issues include back relaxation under DC actuation, hysteresis [35, 36], dehydration in air which alters the stiffness and natural frequency [4, 14, 37], electrolysis [4, 38, 39], non-uniform bending [14], extreme environmental sensitivity (hydration, temperature, humidity level etc. [5, 40, 41]) and loss of mechanical force at larger displacements [16, 38, 42]. All of these issues imply that IPMCs are extremely nonlinear (especially at high inputs and low frequencies [5, 19, 39]) and time varying.

Over a period of operation the highly time-varying nature of the IPMC will cause its response to change unpredictably. This cannot be fully modeled as the variance is due to ion redistribution which is a stochastic process. This random behavior also makes IPMCs very difficult to accurately control. Robust and adaptive control methods must be used to make these systems reliable when operating over a period of time.

Despite the extremely complex nature of IPMCs, modeling has been undertaken to give a relatively accurate representation of the IPMC response in order to aid in the design of mechanical systems and permit performance simulations before implementation into real applications. Models alone, however, are not accurate enough over time to be used to develop controllers simply in simulation, thus online experiments must be used to tune IPMC controllers.

1.3. Biomedical Robotics

Biomedical research encompasses a number of new research areas including biosensing, bioimaging, biomechatronics, biorobotics, bioinformatics and tissue engineering which have been formed through the multidisciplinary cooperation between medical practitioners, scientists as well as engineers from a number of specializations. This new field of biomedical research has resulted in new insights, therapies and solutions to medical problems, leading to advancements in overall healthcare [43].

Biomedical robotics focuses on devices which can be used to mimic and hence replace or augment biological systems as well as external devices which can be used for medical analysis and research. Examples of some biomedical robotic devices are human prostheses, heart compression devices [44], cell micromanipulators [45, 46], active catheters [47, 48], active Braille displays [49], microfluidic pumps [50], polymer drug delivery [51], as well as robotics for medical surgery, orthopedics, musculoskeletal and visceral surgery.

Development of such complex biomedical devices demands novel and highly multidisciplinary approaches to system design.

1.4. Objectives and Scope

The overall aim of this book is to present the development and implementation of biomedical robotic devices with integrated IPMC actuators. A number of fundamental objectives have been identified as crucial in achieving the overall aim of this book and are introduced below.

1.4.1. An electro-mechanical Design Based Model of IPMC Actuators

The first objective is to develop a new comprehensive model for the complete mechanical actuation response of IPMC transducers. The model will be used as a design tool for developing biomedical devices, to enable simulations and design optimization before the actual systems were built. The model will also be used for simulating and hence verifying the proposed control schemes as well as for determining initial controller parameters for the devices before carrying out any experiments.

For the model to be capable of carrying out these tasks it must be devised based on real physical phenomena, which mimics the true transduction mechanisms as shown in Fig. 1.3, in order to be scalable for mechanical design. The model architecture must also be relatively simple so that it can be used for designing and simulating real-time control systems. The model must account for the effect of all external loads and disturbances, which has not been previously achieved. The model will be the foundation building block for developing the biomedical devices presented in this book, giving insight to the actuation performance of IPMC actuators embedded in devices. This will represent a major progression in IPMC modeling, enabling a wide variety of previously unachievable research through its increased capabilities and accuracy.

1.4.2. Design of IPMC Actuated Biomedical Devices

With the exceptionally unique actuation properties of IPMCs, in comparison to traditional actuators, truly novel and inventive approaches must be taken to designing mechanical systems with these embedded transducers. With the aid of the developed design based IPMC model and advanced CAD tools a number of novel devices will be devised, all for distinct biomedical applications. The proposed devices include a bio-inspired compliant stepper motor, artificial muscle finger joint for prosthetics or rehabilitation, a microfluidic pump for drug dispensing, a complete cell micromanipulation system with microtool/gripper and a force compliant surgical robotic tool. These systems all have specific purposes and must exhibit certain properties to successfully achieve their necessary performance. In the design stage these devices will be simulated to confirm their performance before a custom control system is developed and the system is implemented. As well as theoretical simulations, this design stage must also tackle the practicality of actually fabricating the devices and a number of other issues associated with operating them in a real world environment.

1.4.3. Development of Advanced Control Methods for IPMCs

The main motivation for employing control is to reliably guarantee device performance. As IPMCs exhibit highly nonlinear and time-varying properties, as discussed in Section 1.2, advances must be made on previous efforts in literature in order to successfully control their behavior to achieve each application's required operation. Also as each application has a unique performance criterion, it is necessary to tailor a custom controller's properties for each specific application to be successful. No one controller can be used for all IPMC applications and this is why a number of different controllers have been developed. The controllers must all be adaptive to account for the time-varying nature of the IPMCs themselves as well as be adjusted based on real experiments as IPMC dynamics will drift far from any proposed model and even outside the limits for model based robust controllers [52].

To achieve this objective a model-free iterative feedback tuning algorithm has been proposed to tune the system performance. Iterative feedback tuning has its roots in adaptive control theory. Iterative feedback tuning couples a linear controller with an automatic tuner. New iterative feedback tuning algorithms must be developed and implemented to customize the standard tuning algorithm in order to exhibit the desired performance characteristics required for each of the proposed devices; the finger joint requires nonlinear control due to its high displacement, the micro-pump controller must be adapted online during normal operation to maintain a constant flow rate, the microtool/gripper and micromanipulator must be robust to external disturbances and the surgical robotic tool needs accurate force control to ensure compliance. These new algorithms will have superior performance to controllers which have been previously implemented for controlling IPMCs.

1.4.4. Implementation and Testing of the IPMC Actuated Biomedical Devices

The final aim of this book is to share the implementation and testing of the proposed devices and their corresponding controllers to verify system performance in real applications. This validates the entire design process from the IPMC model and mechanical system simulations through to controller performance and device functionality.

Chapter 2

State of the Art: IPMC Modeling, Control and Applications

This chapter summarizes the current state of the art technology in IPMC actuators. First a brief history of the major developments is detailed to give a background of the technological advances since the conception of IPMCs in 1992. After this a survey is presented which focuses on the current modeling techniques, IPMC control systems and also the implementation of IPMCs into both biomedical and other applications. This review does not cover any literature on IPMC manufacturing or sensing, although some current and notable literature in these areas can be found in [33, 52-59] and [20, 21, 60-69] respectively.

2.1. Historic Development

Direct transformation from electrical energy to mechanical work using an ionic polymer was first reported in 1965 by Hamlen *et al.* in [70]. Since then many pioneering researchers have been working on developing ionic gels and studying their response with electric stimulus. The first working IPMC actuators were developed by Oguro and his co-workers in 1992 [71], which consisted of a polymeric gel with plated platinum electrodes and a similar idea was proposed by Shahinpoor in the same year [72]. These IPMC actuators had a higher response and were more robust than the polymer gels at the time. The first reports of an IPMC sensor were by Shahinpoor [72] and Sadeghipour *et al.* [60] in 1992. In the early stages the IPMC was also known as ionic conductive polymer gel film (ICPF), but currently is almost exclusive called IPMC.

After these initial breakthroughs many researchers have been working with IPMC actuators in many areas including, (i) improving their microstructure and composition through manufacturing techniques to enhance the mechanical and electrical responses as well as their robustness and life cycle, (ii) applying IPMCs in a range of laboratory applications, (iii) modeling their response to gain understanding and insight in to the actuation principles and (iv) controlling output response in order to carry out useful tasks.

The first model to describe the response of an IPMC actuator was developed in 1994 by Kanno *et al.* [73] which was a simple linear time-invariant (LTI) empirical model based on curve fitting experimental results. Since then many researchers have developed a number of models based on experimental and analytical approaches all with the intention of revealing some further insight into the response of the IPMC actuators.

For 10 years since their invention all IPMC research work had been on the chemistry, fabrication and modeling to improve the understanding and response. It wasn't until 2001 that the first reported case of feedback control was presented by Mallavarapu *et al.* [36]. Full state feedback was applied, with the gain matrix obtained using linear quadratic regulator (LQR) optimal control techniques. The controller reduced the settling time by a factor of 10 and effectively eliminated the overshoot observed in open-loop response demonstrating the power of implementing close-loop control. After this initial work more advanced controllers are constantly being developed, to this day, in order to improve the performance of IPMC actuators. The advantages and shortfalls of all the current control systems will be examined in Section 2.3.

Applications for IPMCs have been proposed right from the very first report of IPMCs by Shahinpoor in 1992 [72]. Almost all applications have been operated in open-loop with little close-loop control applied to IPMCs in real applications. Most current applications have been confined to the laboratory and very little research has been undertaken in real world conditions.

The next step in IPMC development will be the fusion of all of the current research; modeling the IPMCs to evaluate their performance as actuators and sensors in a specific designed application, implementing custom advanced control techniques for the application and then realizing the entire embedded system in real life. This is the aim of the work presented in this book, to advance the state of art in IPMCs by developing new biomedical applications through fundamental research in IPMC modeling and control.

2.2.   Modeling

Accurate models which describe the behavior of IPMC actuators are essential tools for both understanding the actuation mechanisms as well as for designing systems which incorporate Ipmcs, to allow simulation and evaluation of their performance in the real world. Although there has been a large amount of research into the material behavior and actuation mechanisms of IPMCs, no complete and widely accepted model has been developed to predict the current drawn as well as the mechanical output [5, 74]. This is mainly due to the very complex nonlinear, time-varying and unrepeatable nature of the composite material [38, 75, 76].

There have been quite a number of models proposed in literature and based on their architectures can generally be categorized into three types, black box, grey box and physical models, ranging from the simplest empirical type through to the most complex electro-chemo-mechanical types. Each modeling approach has its distinct advantages and disadvantages in the form of simplicity and level of insight. A summary of the key features and contributions of current IPMC modeling is given in the following sections.

2.2.1.   Black Box Models

Black box models are based solely on system identification of the IPMC material and so no consideration is given to the underlying physical principles. The main advantage of this type of model is that it is easy to develop and can be implemented in a short period of time to give the user some idea of the IPMC response. Engineers and designers working on force and position control of IPMC benders have developed various black box models, a number of examples can be found in [8, 34, 36, 37, 39, 73, 77 -82]. A model which empirically characterizes the IPMC performance with dehydration as a function of time and applied voltage is given in [83]. Although these models are easy to develop their scope is very limited as they are size and sample dependent and so they are not scalable or transferable to different IPMC materials [36, 38]. Due to these reasons black box modeling is not considered in this research.

2.2.2.   Physical Models

Physical or white box models are based on a first principles approach and attempt to model what is believed as the set of underlying physical

31

and chemical transduction mechanisms that produce the resulting electrical (sensing) and/or mechanical (actuating) output. These models have the most structure and scalability, but are the most complex to derive and characterize. A number of researchers have developed such models in [75, 78, 80, 84-92]. Physical models are typically too complex for use in practical applications such as design simulations and for developing real-time controllers.

2.2.3. Grey Box Models

Grey box models offer more insight into the physical transduction of the IPMC in comparison with black box models yet their governing equations are not too complex to solve so they are still very useful in practical applications. Grey box models are therefore a middle ground, where the fundamental model characteristics are derived from real physical principles so the empirically determined parameters have a physical interpretation. These models consist of simple lumped parameter equivalent models to describe each stage of transduction and when coupled together simulate the overall IPMC response. Grey box models are the most useful for engineers as they are geometrically scalable, accurate over a number of different operating conditions and inputs, yet concise and sufficiently uncomplicated to remain practical for engineering design and control. Grey box model architecture will accommodate all the necessary requirements of the model needed for this research and as such grey box models are concentrated on.

The first and most widely accepted grey box model was developed in 1996 by Kanno *et al* [93]. This was a linear dynamic model for ICPF which split the response into an electrical, stress generation and then mechanical stage. This model was extended by Kanno *et al.* [94] into 3 dimensions. Both models use finite element analysis techniques for the mechanical stage. These models were fairly primitive in that they only included a simple electric circuit model and the experimental validation was limited.

DeGennes *et al.* [95] proposed a model, which has the potential to model both the sensing and actuation response, but only the steady-state response is considered. This model was developed further by Paquette *et al.* [96].

In 2002/2003 a 'transducer' model [97-99] was developed by Newbury and Leo to describe both the sensing and actuation properties of IPMCs. This was expanded on in a PhD thesis by Newbury [100].

Although not developed from first principles it does provide some insight to the electro-mechanical behavior and is the most design oriented model that has been proposed. More recently Bonomo *et al.* [38] extended the Newbury model by adding a nonlinear branch into the electrical circuit stage to improve the accuracy of the model.

All the previous grey-box models have fallen short of the requirements for this research. In all of the models, except for Bonomo *et al.* [38], a simple linear RC electric circuit has been used to predict the current draw even though it is commonly accepted that this is a non-linear phenomenon due to electrolysis and hysteresis. In Bonomo *et al.* [38] a simple nonlinearity is added but the hysteresis effect is not specifically dealt with. None of the models have taken into account the clamped section of the IPMC and what effect this will have on the overall actuation response. Also none of the models have been designed to predict or have validated the displacement, velocity, blocking force and force at varying displacements. The force versus displacement relationship is a major issue when designing applications for driving loads as the IPMC loses available force at higher displacements. No model has predicted the large mechanical response with high input voltages i.e. > 3 V. Also no model has taken into account the behavior of the IPMC when driving an external mechanical device or carrying a load. All the grey-box models except Kanno *et al.* [93] use a simple beam model which assumes small tip deflections in relation to the beam length; this is not acceptable for an IPMC actuator.

The model developed and proposed will be presented in Chapter 3 and will solve all the issues that have arisen from the existing grey box models to a truly accurate and useful scalable model for design of IPMC actuated systems.

2.3.   Control

It is widely recognized that simple open-loop inputs cannot accurately regulate the complex IPMC actuation due to the stochastic nature of the electro-chemo-mechanical transduction mechanisms. Richardson *et al.* reported in [37] that open-loop control using a developed empirical model gives extremely slow and inaccurate performance even in ideal conditions. As a result of poor open-loop performance there has been much research into improving IPMC performance through implementing various close-loop control algorithms to regulate the voltage signal being applied, based on the state of the IPMC, in order to accurately achieve a desired mechanical output. This is not a trivial task!

Wide-ranging methods of control have been reported in literature all of which have advantages and disadvantages, usually sacrificing accuracy and/or reliability for simplicity or vice versa. The vast majority of these control methods reported have been implemented to control the tip displacement using a laser sensor or vision system, or force with a precision load cell, to follow step and dynamic reference signals in ideal laboratory environments with no external loads or mechanisms. The following sections give an overview of the current research efforts in controlling IPMC actuation.

2.3.1. Linear Control

A number of researchers have demonstrated the success of classical PID controllers (and variations, P, PI etc.) on IPMC actuators [8, 37, 79, 101-103]. All these studies empirically develop an approximate linear model for the IPMC and use this to calculate the controller gains. Yun and Kim [104] used this method to develop a PID controller and implement it with an integral anti-windup scheme to improve performance. Richardson *et al.* [37] has implemented a linear impedance control algorithm. Optimal control has also been demonstrated using LQR techniques by several researchers [36, 77, 101].

Linear controllers are the simplest algorithms to implement and have been found to improve the open-loop response drastically. They have a reasonable response if the target references are in a small operating range which remains approximately linear. The performance will rapidly degrade with larger target outputs. Also during operation the IPMC system dynamics may drift far from the linear model which was used to tune the gains. The controller gains need to be regularly updated throughout operation as the performance depends on hydration, operating environment, age of the IPMC etc. Simple linear controllers are commonly used as a benchmark to measure the performance of more advanced IPMC controllers against.

2.3.2. Reference Model Based Control

Chen et al. [35] have modeled the quasi-static response of IPMCs and then utilizing an inverse Preisach operator with a linear dynamic model have created an open-loop controller for the hysteresis behavior in an IPMC. This control scheme was expanded on by Zhen *et al.* [34] to account for the creep (or back relaxation) phenomena apparent in IPMCs. In [105] a method to inverse these models in real-time is proposed, building on the previous efforts. These controllers show improved performance over standard open-loop control, but are hindered as the models simply cannot fully address all the characteristics of an IPMC. Due to the highly non repeatable behavior and the time varying dynamic properties of the IPMC as well as any disturbances added to the system, open-loop techniques are not suitable for accurate control even if using an apparently 'accurate' inverse model. A 2-degree-of-freedom (DOF) controller, which uses the inverse of a linear IPMC model for a feed-forward compensator in parallel with a standard PI feedback controller exhibits a more robust performance [106. 107]. A similar controller was developed by Yingfeng and Kam [108] where a frequency weighted feed-forward compensator, found through an inverse model, is coupled with a PI feedback controller.

A Model Reference Integral controller was developed by Tsiakmakis *et al.* [109]. This controller has the ability to settle the entire control system in the presence of variations in the dynamic behavior of the system. It achieves this by forcing the actual dynamic behavior to follow the reference model and hence accurate and consistent tracking controllers can be developed.

2.3.3. Nonlinear Control (Non-adaptive)

A nonlinear controller has been proposed by Lin *et al.* [47] for active control of a catheter system. A parameter dependent transfer function was developed based on the dynamic response of the system for varying voltage inputs. The controller then uses a linear parameter-varying (LPV) approach for the closed loop gains. This controller shows good improvement compared with a standard PID controller for the same system. A fuzzy logic nonlinear controller has been proposed and simulated by Khadivi *et al.* [110]. Tracking performance of these nonlinear controllers has been shown to be higher in comparison to other conventional linear controllers, but as these are non-adaptive they will struggle to cope with the time-varying nature of the IPMC if operating for a long duration.

2.3.4. Robust Control (Non-adaptive)

Three types of robust controllers; H_∞, H_∞ with loop sharing and μ-synthesis were developed by Kang *et al.* [79]. These robust controllers all have improved performance compared with a conventional PID controller, in terms of faster actuation, lower overshoot and smaller voltages. Another H_∞ controller was developed in 2008 by Chen and Tan [111], which showed good tracking performance to a sinusoid based reference signal. A robust controller was developed using quantitative feedback theory (QFT) in [6], this also shows superior position tracking compared to a PID controller. In [112] a nonlinear IPMC model with hysteresis and some injected uncertainties is used to validate an operator-based robust controller. Robust PID force control has been implemented by Sano [113] and using Kharitonov's theorem the robust stability is satisfied.

All these robust controllers show good response in simulations with some perturbations but experimental validation is very scarce. No studies verify the controllers' robustness with the IPMC driving a load or with experimental mechanical disturbances. Also the models used are all time invariant, with perturbations or uncertainties within a certain range, where in reality the fact is that IPMC dynamics do vary far from their initial state and the performance of a static model-based controller can become unacceptable in real experiments. All the robust control methods have only been proven to operate well over a short period of time; the real system dynamics can change so much that they will even shift outside the range the robust controller has been designed for [51].

2.3.5. Adaptive Control

IPMCs are extremely sensitive to environmental changes [39, 40]. Their dynamic response also varies considerably with time suggesting the use of adaptive control schemes for tracking performance of IPMCs for prolonged operation and under varying environmental conditions. A MRAC scheme was presented by Lavu *et al.* [39] that incorporates an intelligent tracking controller, which can adapt the system behavior to allow for noise and variations in relative humidity of between 90 – 98 %. Limited simulation results are presented using a humidity dependent model but no experimental results to verify that the algorithm does actually improve tracking performance over a period of time. Another MRAC is implemented by Brufau-Penella *et al.* [40], where a LTI model is used as the reference. In their work some experimental results are presented to demonstrate the adaptive behavior of the controller but this is over a relatively short period of time, only operating up to 4 minutes in one experiment.

An adaptive neuro-fuzzy controller (ANFC) has been implemented in [114] to show satisfactory results when controlling an IPMC tip position. Extensive training is required to adjust the membership functions in the fuzzy control algorithm. The system has shown better performance than a non-adaptive pure fuzzy controller (PFC). An emotional learning controller using the neuro-fuzzy Takagi-Sugeno-Kang (TSK) method is applied to govern the IPMC dynamics in [115]. A Robust adaptive controller with leakage modification is developed by La and Sheng [116] to control a nonlinear model of an IPMC. Only very limited simulation results are presented with no experimental validation.

It is well known that IPMC dynamics vary far from their initial state over a long period of operation, especially through dehydration when operating in air. The performance of static model-based controllers can then become unacceptable in real experiments and so adaptive controllers are implemented to overcome this. Many reports on adaptive IPMC controllers claim to be capable of adapting to the IPMC response, yet none have presented any convincing experimental results to verify that their controllers can operate for a prolonged time. Guaranteeing accurate operation over a period of time has been one of the major stumbling blocks for implementing IPMCs into real applications.

2.4.   Applications

Since the advent of IPMCs researchers have been proposing possible applications for this smart material. Numerous applications from biomimetics to industrial robotics have been considered attempting to make use of the desirable properties while avoiding the shortcomings. Currently most IPMC applications have been very niche and have remained in the laboratory; none have made it to the real world mostly due to the lack of reliability of the materials without some advanced embedded control system. Most applications have only been conceptualized and not actually fully tested in operation. There has been very limited analysis or experimental examples of research which considers IPMC interaction with driving an external load or mechanism; most simply consider simple cantilever actuation. There are few examples of close-loop control of an IPMC with an application and these have remained operating in laboratory conditions. The following sections describe all the current biological and other IPMC applications that have been found in literature.

2.4.1. Biological

As IPMCs have properties which tie them closely to biological systems much research has been aimed towards biomedical and biomimetics applications. Shahinpoor and his co-workers have manufactured a number of different 'artificial muscle' configurations, for example rod and coil types as well as the common beam muscles [4]. This has led to researchers exploring possibilities in human augmentation for use in wearable exoskeletons or prosthetics to actuate joints such as fingers [4, 117-119] as well as for sensing human joint motion [120, 121]. A prosthetic limb controlled by neurological signals has even been proposed [122]. Configurations of IPMCs to produce linear actuation for a biped walking robot have been proposed [18, 123], this has the potential to be extended to mimic human muscle for prosthesis or eventually even replace a damaged human muscle.

Some medical devices proposed by researchers include an implantable heart compression device using thick IPMC actuator strips to help patients with cardiac problems [15, 43], a micro-catheter with an IPMC tip that can bend to redirect the tube through different paths in the body [47, 123, 124], multidirectional IPMC for microendoscopic ocular surgery using optical fibers [28, 125], active Braille displays for the sight impaired [123] and an active helical stent which can be inserted into a natural passage/conduit in the body to prevent flow constriction [13]. A number of different configurations for microfluidic pumps have been proposed which include a double diaphragm mini-pump [4], single diaphragm actuated micropump [126] and an infusion micropump [127]. A tactile pressure display which gives haptic feedback to a human finger for telemanipulation surgery has been proposed [128].

Natural biological systems are the result of millions of years of evolution and as such are extremely elegant and efficient. For this reason biomimetics, where IPMCs are employed to mimic biology, is a popular application area. Learning from the actuation properties of biological systems can lead to device miniaturization and performance improvement when compared with traditional robotic systems. IPMCs can be manufactured to different geometries and so are good candidates for mimicking a large range of biological systems. As IPMCs can operate in aqueous environments a number of underwater swimming robots have been developed to replicate the bending motion produced by the fins of fish [24, 129, 130] as well as snake-like swimming robots [123, 131, 132], a jellyfish like micro robot [133] and an octopus [4]. IPMCs can also mimic the behavior of bird and insect wings reasonably well and so a number of flapping and flying robots have been developed [4, 134]. An amoeba-like robot, which can change shape to get into small spaces [135] and a wormlike robot [136] have been designed and can be used for exploring hard to reach places like earthquake devastated buildings.

2.4.2. Other IPMC Applications

One of the first major attempts at a commercial application for IPMC bending actuators was a dust wiper, developed by NASA for the Nanorover in the MUSES-C Mars mission [25]. Another application is in vibration suppression of a flexible structural member in [137, 138], where IPMC patches were attached to an aluminum beam and actuated 180 degrees out of phase with the vibration. IPMCs scalability, low voltage requirements and small size makes them ideal for MEMS devices [139]. Several attempts have been made at constructing linear actuators which convert the bending motion of IPMC into linear motion to be used for replacing traditional linear actuators [4, 139]. An inverted pendulum was controlled in an underwater environment in [140, 141]. Thick IPMCs made by hot pressing a number of thin Nafion® layers together to increase the force output have been developed by Bonomo *et al.* [142]. Also a number of robot grippers have been developed, utilizing the bending of two or more actuators to grasp an object between them with enough force for lifting [4, 143-146]. A preliminary study for wireless actuation of IPMCs was recently presented by Soo *et al.* [147].

2.5.   Summary of the State of Art

The majority of existing IPMC research deals with modeling the behavior of and/or controlling an independent IPMC strip in cantilever configuration in laboratory conditions, ignoring real-world issues such as hydration, gravity and the interaction with an external load or mechanism. In addition most of the existing applications have only dealt with specific niche areas and many have only been proposed but not carried out in order to tackle the challenging issues that come with implementation.

In this book a grey-box model, with improved accuracy, that can be used for analyzing different operating modes and conditions with any external loads and disturbances is presented, allowing a much larger range of systems to be simulated. All applications will be developed using the design based IPMC model and with a vision to implementing them into real life situations, with issues such as mechanism dynamics, friction and weight being tackled so these systems could potentially be used as replacements for existing, more bulky devices. Each device will have a controller tailored to its specific requirements whether it is nonlinear for large displacements or robust and adaptive to account for environmental disturbances as well as for driving loads. All the new controllers will be capable of reliably actuating the IPMC over a long period of time to cope with the time-varying dynamics even if they drift far from a nominal IPMC model. The long-term controllability will be validated.

Previously each of the areas of IPMC research and development (modeling, control, application design, analysis and implementation) has been undertaken somewhat independently. No research has gone through a thorough design process by combining all the areas of research to develop a complete solution for a real life application.

This book will present new research that not only plugs a number of gaps in the fundamentals of IPMC modeling, control, design and implementation but demonstrates the importance of designing applications from first principles rather than using an ad-hoc method which has been the common approach taken in current literature.

Chapter 3

A Comprehensive Scalable Model for the Complete Actuation Response of IPMCs

An accurate model describing the behavior of IPMC actuators is an essential tool for engineers designing systems incorporating IPMCs, to allow simulation and evaluation of their performance in the real world. The model must be able to predict the dynamic displacement and force output, and the relationship between them, as well as the current and hence power consumption. The model must be capable of predicting the behavior of the IPMC when driving a load or interacting with the external environment. Also as IPMCs can be tailored to any geometry, it is very important to have a model which is geometrically scalable, this way the appropriate size of IPMC required for a specific application can be found through simulation, before actually fabricating the actuator.

Despite the fact there have been a large number of models proposed in literature, as outlined in Section 2.2, no model meets all the requirements for this research and so a new model has been developed. This model will overcome many of the shortfalls in previous modeling work and hence will result in a more advanced and accurate model for the complete actuation response of IPMC actuators.

Extensive experiments have been undertaken and the results have been used to develop and validate this model. IPMC strips with varying dimensions have been experimented with and results are presented to validate the scalability of the model.

This new model was first proposed at the 2nd International Conference on Smart Materials and Nanotechnology in Engineering [149] and has been published in full in the Smart Materials and Structures Journal [16].

3.1.   Electro-mechanical IPMC Model

The new model developed can accurately predict the full dynamic actuation response of an IPMC actuator, including free displacement and blocked force at varying displacements. An accurate relationship is found between the blocked force and free displacement, as a function of position along the length of the IPMC strip and time. Also because the actuation response, free displacement and blocked force, have been modeled in the time domain, the velocities, accelerations and impulse response can easily be derived.

The model takes an applied voltage input and the resulting mechanical outputs are modeled in the rotational coordinate system to give more practical results for the design and simulation of mechanical devices. The model also allows for the application and interaction of the IPMC with any type of external load or mechanical system. The predicted current draw gives the power consumption of the device which is essential when developing for embedded or remote applications. The model is also geometrically scalable for different sizes of actuator and is designed for operation in air.

The model is particularly intended for biomedical robotics and therefore is designed for low frequencies to DC and for large, up to ±3 V, inputs and the resulting large displacements. It has been shown in [38, 150] that at low frequencies IPMCs are extremely nonlinear and difficult to model due to non-repeatability [100]. The proposed model tackles these issues in order to enable accurate simulations when operating in these states. It has also been shown in [150] that the nonlinearities are majorly dependent on the level of input voltage signal and therefore a number of the model parameters vary with respect to the input voltage, as will be presented in the coming sections. The model includes nonlinear terms to accurately predict the electrolysis characteristics and the hysteresis behavior is also accounted for in this model.

The actuation of the IPMC has been modeled in three stages, mimicking the real physical mechanisms which cause the actuation. A schematic overview of the newly proposed model is shown in Fig. 3.1, where V_{in} is the input voltage, $I(x)$ is the current draw, $\theta(x)$ is the displacement and $\tau(x)$ is the torque, all as a functions of distance along the IPMC beam, x.

Fig. 3.1. Schematic diagram of electro-mechanical IPMC model [16].

The IPMC is modeled in cantilever configuration in air, with one end clamped in copper electrodes, Fig. 3.2. The tip displacement is defined as the angle, θ_T, between the neutral position and the base-tip line. The blocked torque at the tip, τ_T, of the IPMC is the blocked force measured perpendicular to the base-tip line multiplied by the base-tip distance, directly giving the available torque.

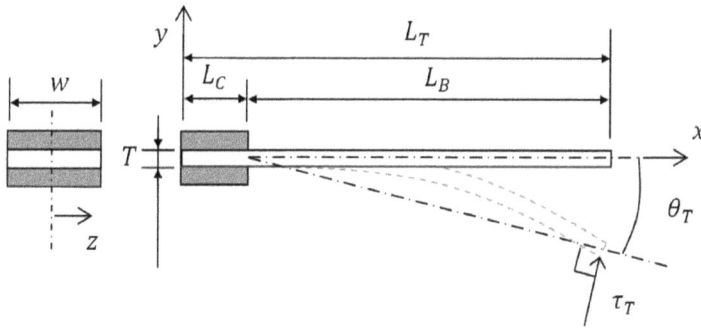

Fig. 3.2. Schematic diagram of IPMC and model parameters.

In order for the model to be completely scalable for different dimensions of the actuator, all parameters throughout the model must be expressed in terms of the IPMC geometry. A list of these parameters is given in Table 3.1. All other model parameters will be introduced and their physical representation explained when they are defined in the text following.

The IPMC model is geometrically split in two parts, as shown in Fig. 3.3, to represent the section 'clamped' by the electrode and the free 'beam' section.

The voltage is assumed to be constant across the width of the IPMC at any point along its length. A lumped parameter nonlinear electric circuit is used to predict the current absorbed by the polymer as a function of length along the IPMC, *I(x)*. Using the model the average current flow can be predicted for the two sections, I_C and I_B.

Table 3.1. Geometric model parameters.

LT	Total length of IPMC
LC	Length of IPMC clamped in electrodes
LB	Length of the free 'beam' section
W	Width of IPMC
T	Thickness of IPMC
θT	Tip angle
τT	Tip Torque

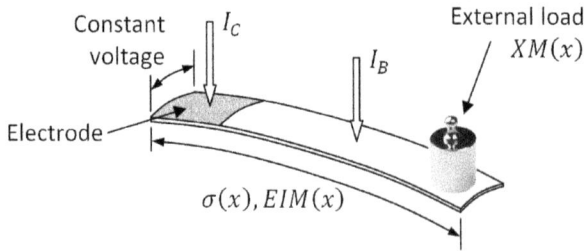

Fig. 3.3. 3D representation of IPMC and model parameters.

The current flow through the thickness and resulting ion flux through the polymer is the major mechanism for actuation [5, 38]. *I(x)* is passed through an electromechanical coupling term (a linear transfer function) to couple the current and resulting ion/water flux profile to the stress induced along the polymer, *σ(x)*, all as a function of length. The stress in the polymer is converted to a bending moment, *EIM(x)*, using the 'flexure formula'. This moment is added with any externally applied moment due to mechanical loads or disturbances, *XM(x)*, and a modified segmented cantilever beam model is implemented to predict the exact elastic curve of the IPMC and resulting mechanical outputs, *θ(x)* and *τ(x)*.

The mechanical circuit has no influence back onto the electrical circuit i.e. is decoupled, as validated in [38] and verified in this model. This assumption can also be justified as the sensing response is orders of magnitude lower than the actuation response. This allows the electrical circuit to be analyzed separately from the mechanical circuit.

3.1.1. Nonlinear Electric Circuit

It has been widely reported that the main mechanism for mechanical actuation of an IPMC is the ion and hence water flux through the polymer. It is therefore extremely important to model the current flow in the IPMC as this can be coupled to the ion flow. Also the ability to accurately predict power consumption is extremely useful for designers.

Due to the electrical characteristics of the material, resistance, capacitance etc., it makes sense to model the current draw using an equivalent electric circuit where the parameters represent material properties. The electrical response is characterized by a dynamic and steady state response. The steady state response is a nonlinear function of the input voltage, which is thought to be mainly due to electrolysis [5, 38]. The dynamic response, can be accurately characterized by a number of RC networks, and has been shown by Newbury in [100] that two RC branches are sufficient to model this response in the IPMC. It is evident that this capacitive dynamic response gives rise to the back relaxation phenomena as well as hysteresis in the polymer, so this circuit is capable of accurately modeling both these behaviors.

The new nonlinear electric circuit is shown in Fig. 3.4. Researchers have proposed models using a similar approach previously [38, 100], also models including a nonlinear capacitance of the IPMC have been presented in [151, 152], but this circuit presents a number of advances over these existing models.

This electrical model divides the IPMC into two separate sections based on its physical configuration, the clamped section and the free beam section. Taking into account the clamped section provides more flexibility when designing the physical configuration and also models the actuation more closely to the real system. Both of these parts of the IPMC will have the same material properties but a different dynamic response as they are geometrically different.

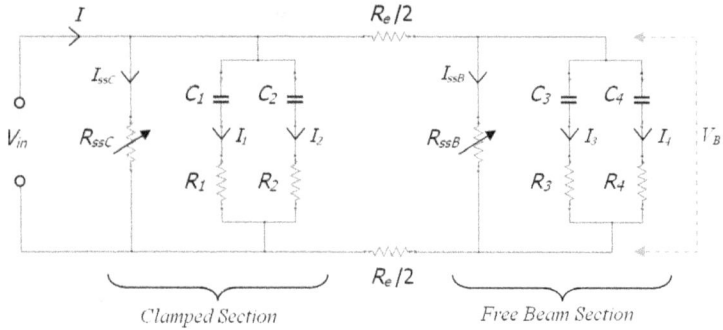

Fig. 3.4. Nonlinear electric circuit diagram used to predict the current drawn by the IPMC [16].

All of the model parameters are based on real physical phenomena in the material and as such are geometrically scalable. This model also gives some insight for users into the internal working mechanisms of the IPMC, yet it is not too complex for design and simulation.

The steady state absorbed current in both sections is modeled using a third-order polynomial dependent on the input voltage signal. Two variable resistors, R_{SSC} and R_{SSB} are used to model this relationship. To account for the change in dynamic response of the IPMC at different input levels, the RC branches are also variable depending on the level of input signal. As a result of characterizing both the steady state and dynamic response of the IPMC as a function of input signal, the current can be accurately predicted in the nonlinear region of operation, i.e. at low frequencies, resulting in accurate predictions of electrolysis, back relaxation and hysteresis inherent in the polymer.

The RC branches 1 and 2 represent the dynamic response of the polymer through the clamped section, while branches 3 and 4 represent the dynamic response through the free section of the IPMC. The two shunt resistors, $R_e/2$, represent the electrode surface resistance and is the average of the ohmic resistance of the surface of the IPMC electrodes. V_B is the average voltage through the polymer thickness in the free beam section, i.e. after half the ohmic loss along the electrodes. The electrode resistance, R_e has been measured experimentally using a four-point probe technique, and using the appropriate correction factor and IPMC dimensions, the R_S (ohm/square) or sheet resistance value can be calculated. R_e can then be expressed in terms of the geometry of the IPMC only and hence can be scaled for different sized actuators using equation (3.1).

$$R_e = R_S \frac{L_B}{w} , \qquad (3.1)$$

R_{SSC} and R_{SSB} account for the nonlinear phenomenon which occurs in the IPMC at very low frequencies and steady state. They also incorporate the equivalent 'through-resistance' of the hydrated polymer membrane. These two material properties cannot be directly experimentally measured so are consequently combined and expressed as an equivalent resistivity, ρ_{SS}. The resistivity, which is dependent on the input voltage, can be found empirically through the steady state relationship between absorbed current and input voltage. The steady state current, I_{SS}, when all dynamic response has died out is approximated as the third-order polynomial in equation (3.2), whose independent variable is input voltage.

$$I_{SS} = aV_{SS}^3 , \qquad (3.2)$$

The equivalent resistance at steady state, R_{SS}, is made up of the resistances R_{SSC} and $(R_{SSB} + R_e)$ in parallel, equation (3.3).

$$R_{SS} = \frac{R_{SSC}(R_{SSB} + R_e)}{R_{SSC} + R_{SSB} + R_e} , \qquad (3.3)$$

The values for R_{SSC} and R_{SSB} can be calculated using $I_{SS} = V_{SS}/R_{SS}$. The resistances are then converted to an equivalent resistivity, ρ_{SS}, using equation (3.4) and (3.5) and can then be scaled and used with all IPMC dimensions. A plot of the real experimental and approximated steady state response for voltages from -4 V to +4 V is given in Fig. 3.5 for a 24 mm long by 10 mm wide by 0.7 mm thick IPMC.

$$R_{SSC} = \rho_{SS} \frac{T}{L_C w} , \qquad (3.4)$$

$$R_{SSB} = \rho_{SS} \frac{T}{L_B w} , \qquad (3.5)$$

47

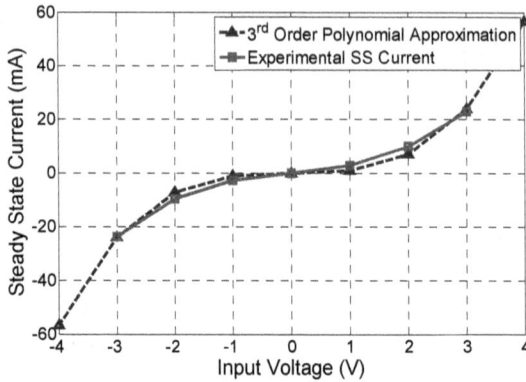

Fig. 3.5. Steady state current draw of IPMC as function of input voltage.

R_1, R_2, R_3 and R_4 represent the resistance against charges flowing through the IPMC that are involved in the dynamic response. They are expressed in terms of the IPMC geometry and an equivalent resistivity in order to enable them to be scaled for different sized actuators. The portion of IPMC in the clamped and free sections have the same material properties and therefore are represented with matching resistivity's ρ_f and ρ_S as in equations (3.6) to (3.9), where ρ_f represents the resistance against fast flowing charges and ρ_S represents the resistance against the slow flowing charges through the polymer material.

$$R_1 = \rho_f \frac{T}{L_C w}, \qquad (3.6)$$

$$R_2 = \rho_S \frac{T}{L_C w}, \qquad (3.7)$$

$$R_3 = \rho_f \frac{T}{L_B w}, \qquad (3.8)$$

$$R_4 = \rho_S \frac{T}{L_B w}, \qquad (3.9)$$

Similarly, capacitors C_1, C_2, C_3 and C_4 govern the time constants for the charges flowing through the IPMC that are involved in the dynamic response. They are expressed in terms of the IPMC geometry and an equivalent permittivity in order to enable them to be scaled. The portions of IPMC in the clamped and free sections have the same material properties and therefore are represented with matching permittivity's ε_f and ε_S as in equations (3.10) to (3.13), where ε_f controls the time constant for the fast flowing charges and ε_s controls the time constant for the slow flowing charges through the polymer material.

$$C_1 = \varepsilon_f \frac{L_C w}{T} \; , \tag{3.10}$$

$$C_2 = \varepsilon_s \frac{L_C w}{T} \; , \tag{3.11}$$

$$C_3 = \varepsilon_f \frac{L_B w}{T} \; , \tag{3.12}$$

$$C_4 = \varepsilon_s \frac{L_B w}{T} \; , \tag{3.13}$$

Now all the parameters for the electric circuit have been defined, the circuit can be analyzed and the current absorbed by the IPMC predicted. Also the average current flow which is responsible for the mechanical response of the IPMC through both the clamped section I_C and the beam section I_B can be found. The current flow through the clamped section, I_C, is the sum of the following branches, see equations (3.14) to (3.16).

$$I_1 = \frac{V_{in} - V_{C1}}{R_1} \; , \tag{3.14}$$

$$I_2 = \frac{V_{in} - V_{C2}}{R_2} \; , \tag{3.15}$$

$$I_{SSC} = \frac{V_{in}}{R_{SSC}} \; , \tag{3.16}$$

The current flowing through the free section, I_B, is the sum of the following branches, see equations (3.17) to (3.19).

$$I_3 = \frac{V_B - V_{C3}}{R_3},$$ (3.17)

$$I_4 = \frac{V_B - V_{C4}}{R_4},$$ (3.18)

$$I_{SSB} = \frac{V_B}{R_{SSB}},$$ (3.19)

where V_{C1}, V_{C2}, V_{C3} and V_{C4} are the voltages across the capacitors C_1, C_2, C_3 and C_4 respectively. The voltage V_B represents the average voltage across the IPMC electrodes and can be calculated using equation (3.20).

$$V_B = V_{in} - (I_3 + I_4 + I_{SSB}) R_e,$$ (3.20)

And substituting in equations (3.17), (3.18) and (3.19) gives,

$$V_B = R_e \left(\frac{V_{in}}{R_e} + \frac{V_{C3}}{R_3} + \frac{V_{C4}}{R_4} \right) \left(1 + \frac{R_e}{R_3} + \frac{R_e}{R_4} + \frac{R_e}{R_{SSB}} \right)^{-1}$$ (3.21)

The dynamics of the entire circuit are governed by the behavior of the capacitors defined by equation (3.22).

$$\dot{V}_{C_i} = \frac{1}{C_i} I_i, \text{ where i} = 1, 2, 3, 4$$ (3.22)

The entire circuit can then be analyzed using suitable numerical algorithms, giving the total current drawn from the IPMC as

$$I = I_1 + I_2 + I_3 + I_4 + I_{SSC} + I_{SSB}$$ (3.23)

The average current density for the two sections of IPMC can be calculated. The current flow through the IPMC will be constant across the width of the IPMC, as the voltage is constant. Current flow through the IPMC as a function of position along the length of the clamped section will also be constant as the voltage supplied by the electrode is constant. The current flow through the IPMC as a function of position along the length of the free section is assumed to vary linearly due to the ohmic loss along the platinum electrodes. The average current flow per unit length in the clamped and free beam sections can be calculated using equations (3.24) and (3.25).

$$I_{C/L} = \frac{I_C}{L_C} \, , \tag{3.24}$$

$$I_{B/L} = \frac{I_B}{L_B} \, , \tag{3.25}$$

The current profile as a function of length can therefore be expressed as shown in equation (3.26) using 'Macaulay Functions' described in equation (3.27).

$$I(x) = I_{C/L} + \frac{2\left(I_{B/L} - I_{C/L}\right)}{L_B} \langle x - L_C \rangle^1 \tag{3.26}$$

The using of Macaulay Function

$$\langle x - a \rangle^n = \begin{cases} 0 & \text{for } x < a \\ (x-a)^n & \text{for } x \geq a \end{cases} \tag{3.27}$$

$$n \geq 0$$

This profile is then passed into the electromechanical coupling term to calculate the conversion of electrical to mechanical energy.

3.1.2. Electromechanical Coupling Transfer Function

It is widely accepted that the conversion of electrical energy to mechanical energy is due to the inner charge/water molecule redistribution [5, 75]. This redistribution gives rise to a number of microscopic actuation mechanisms that induce the macroscopic deformations of the IPMC [75]. These include (i) an increase in electrostatic repulsive force at the boundary layers; (ii) a decrease of the electrical permittivity at the cathode and an increase at the anode; (iii) an increase of osmotic pressure at the cathode and a decrease at the anode; and (iv) polymeric swelling in the cathodic region [75].

All these mechanisms introduce much complexity to the model and they are still not fully understood. The ion redistribution process is stochastic in nature and is responsible for much of the unrepeatability of IPMC actuation. As a result, this model attempts to make a simplification to insure the model is more practical yet still remains realistic. In this model the electric current flow at any point along the length of the beam is linearly coupled to the ion/water flow and hence to a longitudinal induced stress of the IPMC beam at that point. This is physically interpreted as the amount of mass of water that flows through the thickness of the beam is directly proportional to the amount of swelling and stress in one side of the IPMC. A number of different forms for the linear electromechanical coupling transfer function *CEM(s)* were considered and tested. Based on these tests and work in [38] by Bonomo *et al.*, the most accurate response was achieved with the form shown in equation (3.28), which includes one zero and two poles.

$$C_{EM}\left(s\right) = k\frac{s+Z}{s^2 + P_1 s + P_2},$$ (3.28)

The values of *K, Z, P_1* and *P_2* are found empirically. The stress σ as a function of length along the IPMC can be calculated by

$$\sigma\left(x\right) = C_{EM}(s)I(x) ,$$ (3.29)

3.1.3. Mechanical Beam Model

The stress generated along the length of the IPMC, as a result of a voltage input, can be converted to a bending moment, or electrically induced moment (*EIM*) using the 'flexure formula', $\sigma = \dfrac{My}{I}$, where y is the distance from the neutral, x is axis in the y direction and I is the moment of inertia about the neutral axis. It has been reported in [80, 100] that the electromechanical conversions occur at the interface between the electrode and the polymer membrane, using this fact, y is taken as $T/2$.

Now the *EIM* can be calculated as a function of length in the Laplace domain, $EIM\left(x\right) = \dfrac{2\sigma\left(x\right)I}{T}$, and using the Macaulay Function, the *EIM* can be expressed at any point in time by

$$EIM(x) = M_C + \frac{2(M_B - M_C)}{L_B}\langle x - L_C \rangle^1 \text{ where } M_C \text{ is the moment induced}$$

in the clamped section and M_B is the average moment induced in the free beam section, acting at $x = L_C + L_B/2$.

The *EIM(x)* and all other moments which are induced in the IPMC beam as a result of externally applied forces or loads, *XM(x)*, are plotted on a bending moment diagram as shown in Fig 3.6, to calculate the resultant bending moment. In this way this model can accommodate for any external force/moment or load that acts anywhere along the length of the IPMC. This makes the model extremely useful in mechanical design. Blocked force is a specific case where a point load is applied at the tip. The blocked torque can be calculated as the point load required to maintain zero displacement, multiplied by the moment arm, L_B.

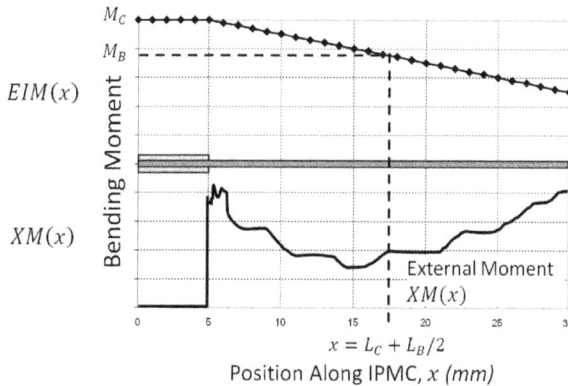

Fig. 3.6. Bending moment diagram of IPMC.

The elastic curve for a beam can be expressed mathematically as, $v = f(x)$, where v is the linear displacement in the y axis and x is the position along the x axis. The nonlinear second order differential equation, $\dfrac{M}{EI} = \dfrac{d^2v/dx^2}{\left[1 + \left(dv/dx\right)^2\right]^{3/2}}$, is used to relate the bending moment to

the beam slope, dv/dx, and the displacement, v. *EI* is the product of the modulus of elasticity and moment of inertia of the IPMC and is known

53

as the flexural rigidity. The solution to this is known as the 'elastica' and gives the exact shape of the elastic curve. To find the elastica however requires the use of higher order mathematics and would over complicate the model. In many applications, in order to overcome this, the above relationship is commonly approximated to $\dfrac{M}{EI} = \dfrac{d^2v}{dx^2}$ by assuming a shallow curve. By ensuring $\dfrac{dv}{dx}$ remains very small, its square will be negligible compared with unity, justifying this assumption [153].

However when modeling the IPMC with large inputs, this assumption will not hold true, and therefore will not give an accurate representation of the true displacement of the beam. Also using this method, as has been done in previous models [38, 100, 111], will only give a linear displacement as a function of the length and not the actual elastic curve and bending displacement. This new IPMC model needs to be accurate for large displacements in order to be useful for the intended applications, therefore another approach is needed.

To overcome this issue, the beam has been 'segmented' into smaller pieces along its length. Providing the segments are small enough, they can be analyzed individually and the shallow curve assumption will hold true. Each segment will have its own elastic curve and when put together will make up the entire elastic curve of the IPMC beam. Using this method will then allow the true shape of the bending actuator to be found, an angular displacement and not a simple linear approximation as in previous models [38, 100, 111]. An example of a typical curve of an IPMC with 30 mm free length is shown in Fig. 3.7(a), with segment length of 1mm. The displacements are more than an order of magnitude smaller than the length of segment, which ensures that the shallow curve assumption holds true.

At the base of each individual segment the initial conditions for the slope and displacement are the tip values of the previous segment:

$$\left.\frac{dv}{dx}\right|_{x=0} = \left.\frac{dv}{dx}\right|_{prev} \text{ and } v\big|_{x=0} = v\big|_{prev}$$

The constraints at the base of the free section of IPMC are consistent with a cantilever beam, i.e. at $x = 0$, $\dfrac{dv}{dx} = 0$ and $v = 0$. Using these constraints permits simulation of the true shape of the entire actuator when the sections are combined as in Fig. 3.7 (b).

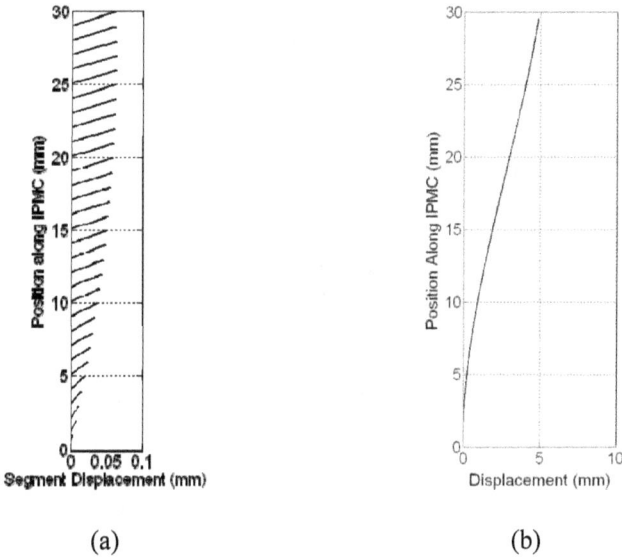

(a)   (b)

Fig. 3.7. Typical displacements for an IPMC with 30 mm free section and segment length of 1 mm. (a) Individual segments and (b) combined elastic curve for IPMC.

Conditions for calculating the free displacement and hence tip angle θ_T are, $XM(x) = M_C - M_C \langle x - L_C \rangle^0$, which specifies that there is no external force applied to the actuator.

Conditions for blocked force are

$$XM(x) = M_C - M_C \langle x - L_C \rangle^0 - P\left(L_B \langle x - L_C \rangle^0 - \langle x - L_C \rangle^1 \right),$$ which represents an equivalent point load at the tip.

The blocked force can be simulated at any displacement by adding the condition for the tip displacement and because the exact distance from the base can be found, the blocked torque can be accurately modeled.

The IPMC material is a non-perfect elastic material [76, 154] and is more accurately modeled using a viscoelastic Young's modulus. This can be easily accommodated using a Golla-Hughes-McTavish (GHM)

55

parameter model as in [38, 100]. This takes into consideration the resonance which makes the model accurate as the IPMC approaches higher frequencies. It has been shown in [100] that this does not play a significant role when below approximately 20Hz, which is out of the range of the applications intended for this model, so has not been included for the parameter identification and verification of this model.

3.2.   Parameter Identification and Results

All the parameters were identified using a Nafion® based IPMC actuator 24 mm long, 10 mm wide and an average thickness of 0.7 mm, with platinum coated electrodes. The clamped length was 4 mm. It has been shown in [119] that the IPMC sample must be hydrated for a minimum of 30 minutes between each test to ensure that the sample has become fully hydrated, so this is the procedure that was followed. The model was developed based solely on the blocked force results for the IPMC and then the free displacement, and force at varying displacements are used to verify that the model is accurate for the full actuation response of the IPMC.

The electrical parameter R_S was measured experimentally using a four-point probe technique and found to be 21.12 Ω. Next, the steady state current for different input voltages from -3 V to +3 V was measured and using a third-order polynomial fit, the coefficient was found to be $a = 8.8403 \times 10^{-4}$ and the corresponding value for ρ_{SS} was then determined.

To identify the parameters associated with the dynamic electrical and mechanical response of the IPMC step voltages from -3 V to +3 V were applied for 60 seconds, to ensure the IPMC reaches steady state, and the current and corresponding blocked force were measured. A 0.1 Ω current sensing resistor was used to measure the absorbed current and a SS-2 Precision Force Sensor (range ± 30 gf) was used to measure blocked force. The force sensor was always orientated perpendicular to the base-tip line when measuring force at the varying angular displacements, so a blocked torque could directly be calculated.

The electrical circuit model is simulated for 60 seconds (note the electrical and mechanical models can be analyzed independently) with the measured values of R_{SSC}, R_{SSB} and R_e and nominal values for the parameters associated with the dynamic response ρ_f, ε_f, ρ_S and ε_S. The correlation between the simulated absorbed current and the experimental absorbed current is calculated using a least squares cost function in equation (3.30), where the values at each time step, i to n are compared. To optimize the values of the four dynamic parameters an algorithm was developed which iterates through, running a 60 second simulation, calculating the error and then updating the parameters, in order to obtain optimal parameter values. The cost function, J in equation (3.30) is minimized using the simplex search method of [155].

$$J = \sqrt{\frac{\sum_{i=1}^{n}\left(\text{Experimental data-Simulated data}\right)^2}{\sum_{i=1}^{n}\left(\text{Experimental data}\right)^2}}, \qquad (3.30)$$

This is repeated to find the optimal parameter values for 1 V, 2 V and 3 V inputs. It can be seen in Fig. 3.8 that there is a clear linear relationship between all the electrical parameters and the input voltage, demonstrating the dependence of the dynamic behavior of the circuit on the input voltage level. The parameters are given in equations (3.31) to (3.34), as a function of input voltage. These provide the best fit over the desired operating range of the model, ±3 V. It can be seen from Fig. 3.8 (a) and (b) that the resistivities will be zero at approximately 3.5 V, so if the model was to be extended beyond this then more experimental data would need to be considered and the trend adjusted accordingly. Throughout this research voltages are limited to saturate at 3 V.

An interesting observation is that the two parameters governing the fast response are varying slowly, where the parameters governing the slow response are more dependent on the voltage input level. This shows that there is more variability at the low frequencies and less at higher frequencies as shown in [150]. This is a good validation for the modeling approach used here.

$$\rho_S = -4.3983|V_{in}| + 15.446, \qquad (3.31)$$

$$\rho_f = -1.2561|V_{in}| + 4.4083, \qquad (3.32)$$

$$\varepsilon_S = 0.4420|V_{in}| + 0.3993, \qquad (3.33)$$

$$\varepsilon_f = 0.1981|V_{in}| + 0.5078, \qquad (3.34)$$

The modulus of elasticity of the composite material cannot be calculated accurately based on theoretical material properties due to imperfect manufacturing techniques, so was measured experimentally.

A force sensor was used to measure blocked force at varying displacements, with $V_{in} = 0$ V, to measure a passive stiffness, then this was scaled to give a Modulus of Elasticity of 0.1757 GPa. The moment of inertia was calculated using the IPMC geometry as 0.2858×10^{-12} m^4.

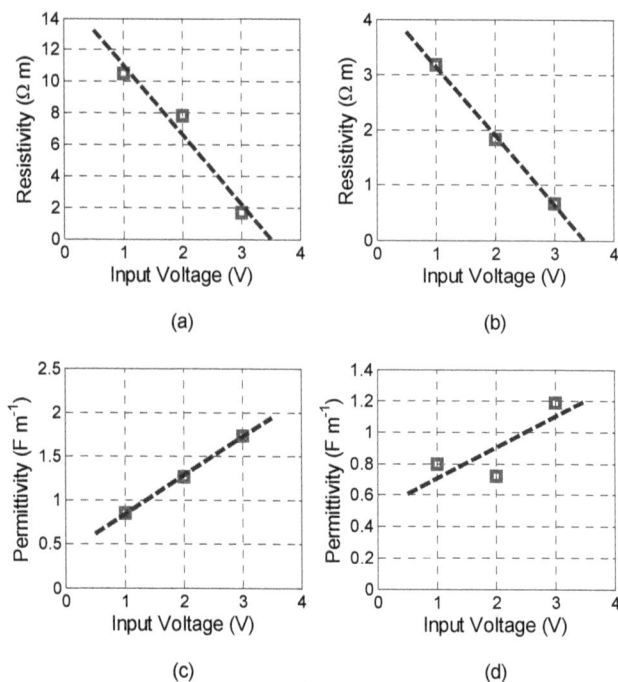

Fig. 3.8. Optimal values for (a) ρ_s, (b) ρ_f, (c) ε_s and (d) ε_f for different input voltages.

The mechanical beam model is now fully developed and constraining the mechanical tip displacement to zero enables the *EIM* for the blocked force condition to be calculated. Once the *EIM* is found, equations (3.28) and (3.29) can be used to optimize the electromechanical coupling transfer function $C_{EM}(s)$. The same process which was used to optimize the electrical parameters is used to optimize the $C_{EM}(s)$ parameter values. Optimal values for 1 V, 2 V and 3 V which minimize the cost function J in equation (3.30) are found and similarly a clear linear relationship was established, demonstrating again the dependence of the IPMC actuation on the input voltage level. The final optimal values are presented in equations (3.35) to (3.38).

$$K = -9434|V_{in}| + 74869, \qquad (3.35)$$

$$Z = -0.1431|V_{in}| + 4.3595, \qquad (3.36)$$

$$P_1 = -2.818|V_{in}| + 7.7198, \qquad (3.37)$$

$$P_2 = -0.0321|V_{in}| + 0.4499, \qquad (3.38)$$

The nonlinear model is now complete with all parameters calculated and presented in terms of input voltage level and actuator geometry so the model is completely scalable and can be used for mechanical design and optimization.

The proposed model with the parameters identified above is simulated for 60 seconds and both absorbed current and blocked force are compared with actual measured values to evaluate their correlation. Fig. 3.9 (a), (b) and (c) plot the simulated and actual experimental results for the current draw for the 24 × 10 × 0.7 mm actuator at 1 V, 2 V and 3 V respectively. The plots clearly show the correlation between the simulated current draw and the actual measured current draw. It can be seen that the peak current draw, the dynamic decay and the steady state values can all be accurately predicted using the model.

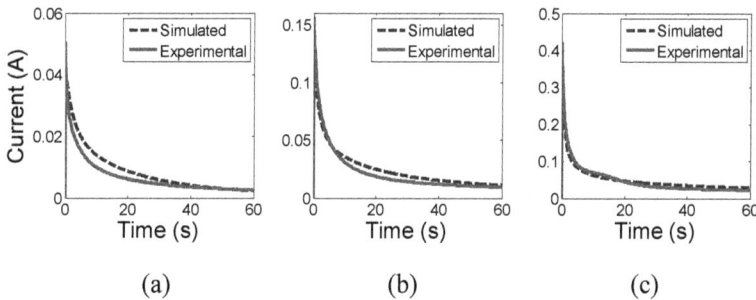

(a)   (b)   (c)

Fig. 3.9. Experimental and simulated current draw for (a) 1 V, (b) 2 V and (c) 3 V step inputs.

Fig. 3.10 plots the corresponding simulated and actual measured blocked force for 1 V-3 V inputs at zero displacement. Again it can be seen that there is a good agreement between the simulated and experimental results. The 2 V experiment is lower than the simulated, but with the unrepeatable nature of the IPMC, some deviations can be expected and are acceptable.

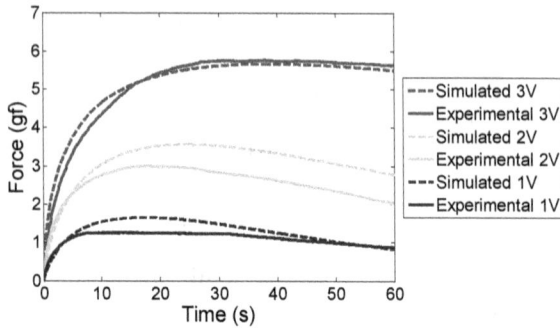

Fig. 3.10. Experimental and simulated blocked force
for 1, 2 and 3 V step input.

Now the model has been developed, it must be tested to confirm the model is accurate for the free displacement as well as the blocked force. A Banner LG10A65PU laser sensor with a 3 μm resolution was used to measure the displacement of the IPMC with no external load applied. It was setup as per Fig. 3.11, and measures the linear displacement at a distance L_m from the base.

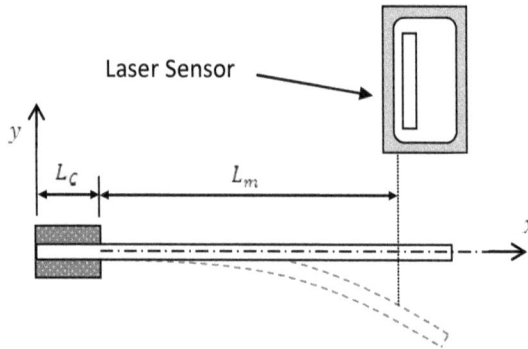

Fig. 3.11. Schematic of experimental setup for measuring free displacement.

The results of the experiments are shown in Fig. 3.12 for 1 - 3 V at L_m = 18 mm. The model is then simulated and the corresponding linear IPMC displacement at a distance 18 mm from the base is calculated. The results of this simulation are also plotted in Fig. 3.12, and show a good match to the actual measured displacements. This illustrates that the model is indeed accurate for both the force and displacement.

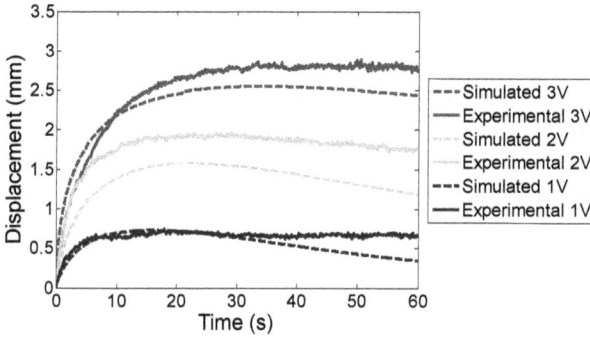

Fig. 3.12. Experimental and simulated linear displacement at $L_m = 18$ mm for a 1, 2 and 3V step input.

It can be seen that the model captures both the nonlinear steady state characteristics, after the system has been left for a long time to settle, as well as the fast dynamic response of the IPMC at a large range of voltage inputs (up to 3 V). The model correctly predicts the back relaxation phenomena and can therefore also model the hysteresis in the material. With the addition of a viscoelastic approximation for Young's modulus in the mechanical model, the simulation will remain accurate from the lower frequencies up to and above the resonance of the actuator.

Fig. 3.13 shows the elastic curve of the IPMC actuator after 30 seconds of a 1, 2 and 3 V step input. This demonstrates the ability to predict the exact shape of the response under varying inputs and also shows the versatility of the actuator model for different applications.

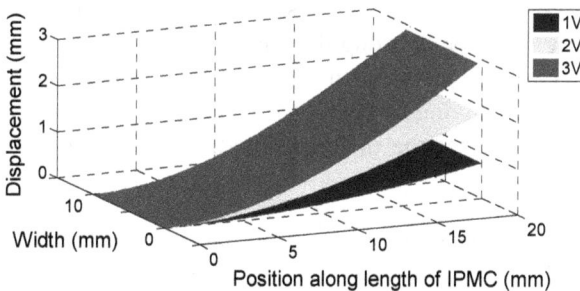

Fig. 3.13. Simulated free displacement response of a 24 mm long (20 mm free section), 10 mm wide IPMC actuator under 1, 2 and 3V step inputs after 30 seconds [16].

Fig. 3.14 shows the result of a free displacement simulation of the 20 mm free length IPMC. The displacement is plotted as a function of the position along the IPMC from the clamped section and of time (blue indicates low displacement and red indicates high displacement). This demonstrates the capability to predict the displacement at any point along the IPMC and at any time, which is very useful information for designing mechanical systems and simulating their performance.

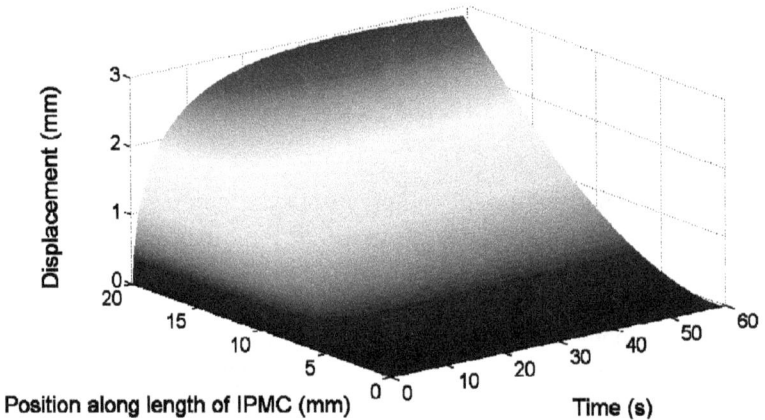

Fig. 3.14. Simulated free displacement response of a 24 mm long (20 mm free section), 10 mm wide IPMC actuator as a function of position along IPMC and time.

3.3. Model Validation

In order to be confident to use this model in real life, under varying conditions, it must be thoroughly validated. Although all the model parameters have a physical meaning, some have been evaluated empirically based on one sample, so they must be corroborated to show they are acceptable for the different conditions that the model will be used for.

The equivalent electric circuit has been designed to accurately account for the nonlinear response at low frequencies as well as the back relaxation and hysteresis effects. The model has shown previously to predict the back relaxation and nonlinear effects for step inputs. In order to validate that this model can predict the hysteresis exhibited by IPMCs [4, 35, 63] the model was simulated for the 24 mm long IPMC, with a 3 V amplitude sinusoid wave of 0.2, 0.1 and 0.025 Hz, over 60 seconds. The results are shown in Fig. 3.15. It can be seen that there is an obvious hysteresis loop as the current draw and tip displacement follow a different path when the voltage is increasing to when it is decreasing. The hysteresis loop predicted by the model is caused by the transient behavior of the variable RC branches in the electrical model. The level of hysteresis is therefore dependent on the input voltage and frequency. This is shown in Fig. 3.15 by the different paths followed for the different frequencies simulated, which is also observed in the real system. It is also clear that the simulated data indeed captures the nonlinearity.

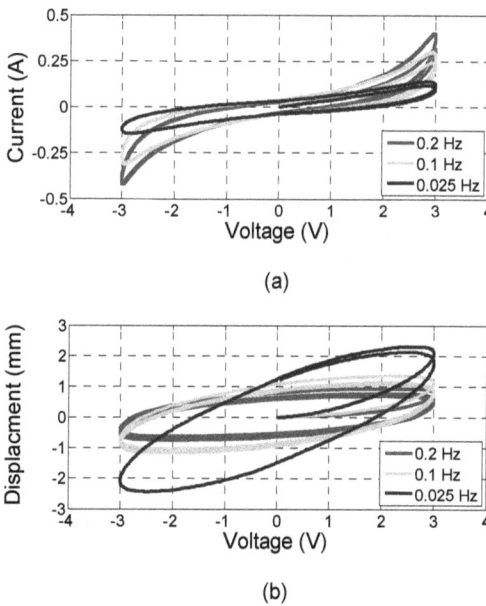

(a)

(b)

Fig. 3.15. Simulated (a) current draw and (b) tip displacement
vs. input voltage for a ±3 V sinusoid wave input of 0.2, 0.1
and 0.025 Hz over 60 seconds [16].

Another claim that must be validated is that the model is accurate for both displacement and force, and the relationship between them. In order to validate this, the blocked force has been measured at a number of different displacements. This will then demonstrate the model's ability to predict the force as a function of displacement as well as the free displacement and velocities of the IPMC, confirming the model is indeed accurate for the complete actuation response of the IPMC.

Fig. 3.16 shows the experimental passive blocked force (0 V input) at varying displacements and the peak blocked force for -3 to +3 V inputs at each displacement. The model is then simulated for the same conditions. The results plotted show the close agreement between the model and the actual measurements.

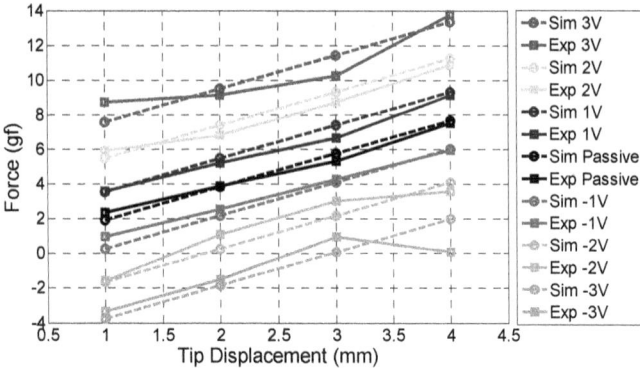

Fig. 3.16. Experimental and simulated peak blocked force at varying tip displacements for -3 to +3 V step inputs [16].

In order to directly compare the dynamic blocked force response at different displacements, the blocked force at 1, 2, 3 and 4 mm tip displacements, with the passive stiffness removed, are plotted. Fig. 3.17 (a), (b), (c) and (d) show the simulated and measured results plotted against time. It can be seen that as the IPMC moves further away from the equilibrium point, the available force remains the same as it is with zero displacement. This fact can be used in designing mechanisms to maximize the force output of the IPMC by controlling the angular operating range of the actuator. So for example you can utilize the passive stiffness plus the electrically induced torque to enhance the maximum force output of the IPMC.

These results also interestingly show that the dynamic behavior of the IPMC does not change as the IPMC bends further away for the equilibrium point. This is an extremely useful conclusion to recognize, and confirms that both the steady state and dynamic response of the actuator due to the application of an electrical input remains the same no matter the displacement from the equilibrium position. This also proves that the response has been correctly modeled for all displacements.

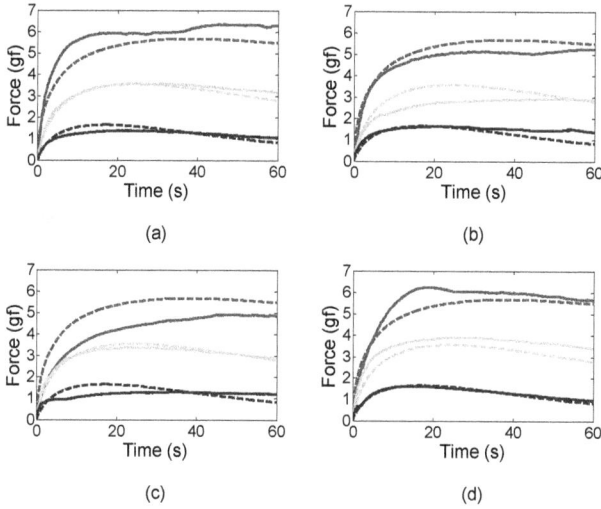

Fig. 3.17. Experimental and simulated blocked force for 1, 2 and 3 V step input at (a) 1mm, (b) 2mm, (c) 3mm and (d) 4mm tip displacements [16].

Fig. 3.18 plots the peak force from -3 to +3 V against tip displacement, with the passive stiffness removed. This plot indicates that at displacements up to 4 mm (or 11°) the electrically induced peak available force, in both positive and negative directions, remains relatively constant at the same value as at zero initial displacement.

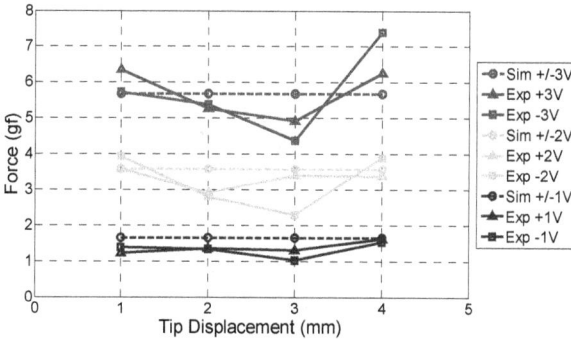

Fig. 3.18. Experimental and simulated peak blocked force at varying tip displacements for -3 to +3 V step inputs, with passive stiffness removed [16].

As mentioned above, all parameters have been developed with a physical interpretation and are expressed in terms of IPMC geometry to

65

ensure scalability. In order to be able to confidently use this model for a complete mechanical system design, its geometric scalability must be validated for both the blocked force and displacements. In the laboratory, experiments were carried out using two different available samples to validate this assumption. Both samples were based on Nafion®-platinum material, one $30 \times 10 \times 0.7$ mm and another $35 \times 10 \times 0.7$ mm actuator, both with 5 mm clamped section. The sets of data were obtained in the same conditions as the original experiments, hydrated polymer in air. Fig. 3.19 and 3.20 compare the results of the experimental data and the model predictions. The free displacements have been measured at $L_m = 23$ mm and 28 mm for the 30 mm and 35 mm long actuators respectively.

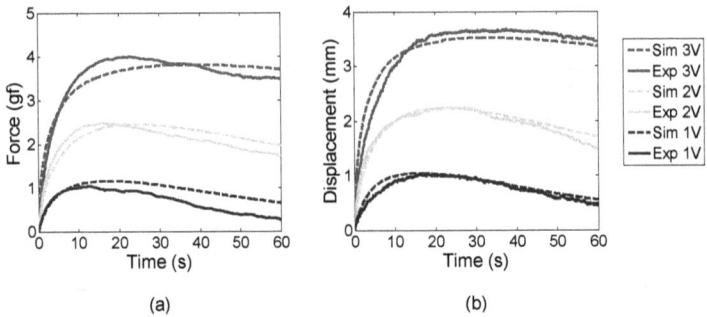

Fig. 3.19. Experimental and simulated (a) blocked force and (b) linear displacement at $L_m = 23$ mm for a 1,2 and 3 V step input of a 30 mm long IPMC.

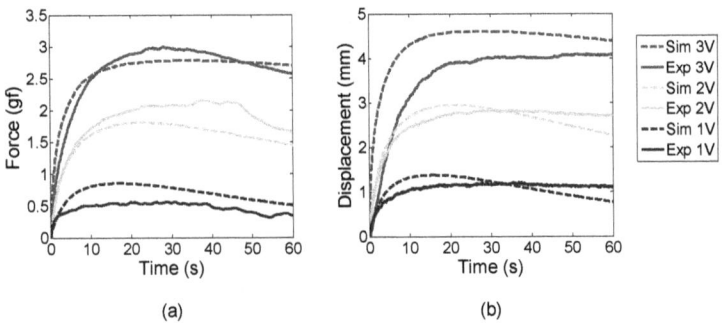

Fig. 3.20. Experimental and simulated (a) blocked force and (b) linear displacement at $L_m = 28$ mm for a 1, 2 and 3V step input of a 35 mm long IPMC[16].

66

These plots clearly show that the complete actuation response, force and displacement, in both the dynamic and steady state range, of an IPMC actuator can be accurately determined for different geometries of actuator. Although there are some deviations, it can be seen here that both the dynamic and steady state response can be accurately simulated for force and displacement, under 1, 2 and 3 V inputs.

Chapter 4

Bio-inspired Compliant IPMC Stepper Motor

The first application to be presented in this book is an IPMC actuated stepper motor. The main reasons for this choice are that many devices require rotary actuation, both in biomedical robotics and other industrial applications, as well the simple open-loop architecture required to control stepper motors will make this system simpler to implement than a full close-loop controlled device. The motor has a design which is influenced by biological systems, with the view to miniaturization and use in biomedical robotics where the force output requirements are low, but other advantages of IPMCs are important, such as weight and power availability for remote and embedded systems. Some target applications may include driving micromanipulators and micro-grippers as well as implantable micropumps and position controlled joints as the motor can operate in air as well as fluidic and cellular environments.

The advantages of the IPMC stepper motor for biomedical devices are biocompatibility, compliance and back-drivability, simplicity and scalability. Advantages of the IPMC stepper motor in comparison to traditional stepper motor include, mechanical compliance, size, scalability, lightweight design and low power consumption.

This stepper motor demonstrates an innovative mechatronics design process for a complete system with integrated IPMC actuators. The motor is developed utilizing the new model for IPMC actuators, in Chapter 3, incorporated with a complete mechanical model of the motor. The entire system is simulated, an appropriate size IPMC strip chosen to achieve the required motor specifications and its performance is then verified. The system has been built and the experimental results validated to show that the motor works as simulated and can indeed achieve continuous 360° rotation, similar to conventional motors. This work was published by McDaid et al. [156].

4.1.   Stepper Motor Mechanical Design

The conventional stepper motor is a brushless, synchronous electric motor which divides a full rotation of the motor into a number of 'steps'. A stepper motor configuration shown in Fig. 4.1 was chosen for achieving rotary motion using IPMCs for a number of reasons, including simple design and working mechanism to convert IPMC bending to rotary motion, very low contact area between IPMC and device resulting in low friction, good holding torque, mechanical compliance and also the ability to use open-loop control architecture. This device converts the bending actuation of IPMCs into a continuous rotational motion of the motor.

The IPMC stepper motor works by sending a voltage sequence to the IPMC actuators which will cause them to bend into contact with the pins attached to the motor shaft, this applies forces to the pins which will then result in controlled rotary motion. The motion, similar to a traditional stepper motor, will be in discrete steps. The pins are placed on a top and bottom layer, each corresponding to one IPMC. The pins on each layer have 90° separation from each other and the top pins are 45° out of phase from the bottom pins. The stand is adjustable for accommodating different size IPMC strips as required. A pulsed input voltage of ±3 V is used to actuate the motor, where each pulse or step will correspond to a 45° rotation.

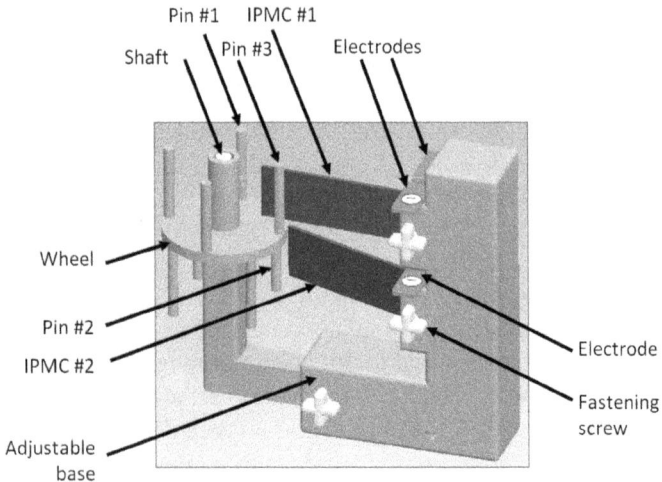

Fig. 4.1. CAD model of the proposed stepper motor, with two IPMCs [157].

This design is based on a biological 'fin' type design where a number of fins are used to produce a rotational motion. As the IPMC fins are naturally compliant the motor is intrinsically compliant, mimicking a biological system. This design can easily be extended to use more IPMCs to provide more driving torque; the two-IPMC system has been implemented first for validation of the design.

The mechanical compliance of the motor is dependent on the stiffness of the IPMCs and the number of IPMCs in contact with the pins. In the two-IPMC system only one IPMC will be in contact at any time and so the mechanical compliance, which is defined as the inverse of mechanical stiffness, of the motor can be described using equation (4.1).

$$s_{motor} = \frac{1}{k_{motor}} = \frac{\theta_{motor}}{\tau_{motor}} \,, \tag{4.1}$$

where s_{motor} is the passive compliance and k_{motor} is the passive mechanical stiffness of the motor. τ_{motor} is the holding torque to an external load or disturbance which results in a shaft angular rotation of θ_{motor}. The passive compliance in this system is physically interpreted as the angle the motor shaft will rotate as a result of an externally applied torque. As the IPMC has a natural passive compliance, the motor is intrinsically compliant, but still has holding torque. This property makes the motor very useful for many applications as it can hold a load, but will not damage a system as it allows for a certain level of back-drivability. The compliance and holding torque are related through equation (4.1); the more compliant the motor is the less available holding torque the motor has and vice versa. These two motor properties are defined by the IPMC material properties and the IPMC and motor geometry. The passive compliance of the motor is described in equation (4.2) where L is the free length of IPMC to the motor pin, E is the elastic modulus of the IPMC, I is the IPMC moment of inertia and r is the radius of the motor wheel. This equation is extremely useful for mechanical design as the compliance of the system required for a specific application can easily be achieved by selecting an IPMC with appropriate material properties and then adjusting the geometry of the IPMC and/or motor wheel.

$$s_{motor} = \frac{\theta_{motor}}{\tau_{motor}} = \frac{L^3}{3EIr^2} \tag{4.2}$$

71

The advantages of this intrinsically compliant motor design over implementing an actively compliant motor drive are the compliance is guaranteed throughout operation, there is no need for complex control algorithms, safety achieved without loss of positioning performance and no expensive force/torque feedback sensors are needed.

4.2. Model Integration and Simulation

The stepper motor has been designed using CAD tools and the properties of the system are then converted to a mathematical model which fully describes the dynamics of the device. The motor friction is added to the system model, using a standard Coulomb and viscous friction model, to make a realistic simulation and see if the IPMC can actually move the motor shaft. The motor friction force was found experimentally to be 0.27 gf for static and 0.21 gf for dynamic motion. Using these forces the friction coefficients were calculated and input to create an entire system model which can accurately represent the real life situation. The complete mechanical model of the motor is integrated with the IPMC actuator model developed in Chapter 3; hence simulations and performance analysis were carried out.

Different lengths of a 700 μm thick Nafion® based IPMC with gold plated electrodes were simulated with a clamped length 5 mm. An IPMC cut from this material which had a length of 35 mm and width of 10 mm was found to give the desired performance with respect to the force required to actuate the motor and deflection to ensure a full step could be achieved. The simulated performance of the system is shown in Fig. 4.2. It can be seen that there is a pause in the operation of the motor between steps. This is necessary in the design with two IPMCs to avoid the motor pins clashing with the IPMC when it is returning to its home position.

Designing the system in simulation first has allowed the system to be extensively tested and the performance verified before the prototype was built. This demonstrates the usefulness of the developed model for designing IPMC actuated mechanisms.

Table 4.1 shows the calculated motor parameters of the final design with the selected IPMCs and the mechanical design as in Fig. 4.1, selected through the model simulations.

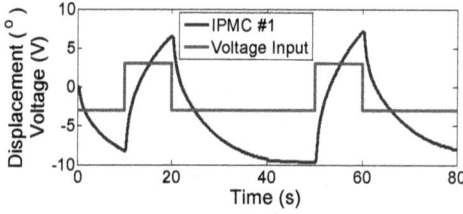

(a)

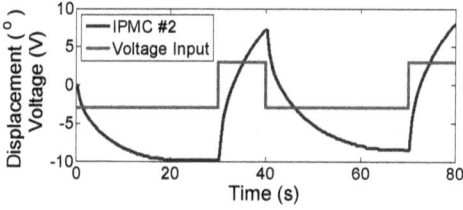

(b)

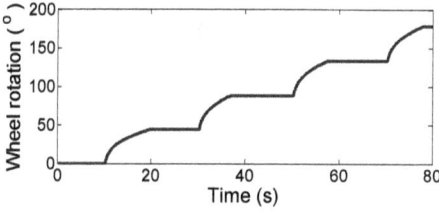

(c)

Fig. 4.2. Input voltage and displacement for (a) IPMC 1 and (b) IPMC 2 and (c) motor shaft displacement [156].

Table 4.1. IPMC motor parameters.

IPMC length	30 mm
IPMC width	10 mm
IPMC thickness	700 μm
Moment of inertia	2.86×10^{-13} m^4
Elastic modulus	0.1757 GPa
Length of IPMC to pin	25 mm
Wheel radius	12 mm
Motor compliance	720 rad/Nm
Holding torque at 5°	1.212×10^{-4} Nm (1.029 gf at motor pins)

4.3.   Experimental Validation

The actual stepper motor has been rapid prototyped from ABS material, shown in Fig. 4.3, and experiments are undertaken to test the actual performance and verify all the simulations. Fig. 4.4 shows the actual measured tip displacements of the IPMCs when they are being actuated in order to move the motor. They are measured using 2 Banner LG10A65PU laser sensors with a 3 μm resolution.

Fig. 4.3. Rapid prototyped stepper motor [156].

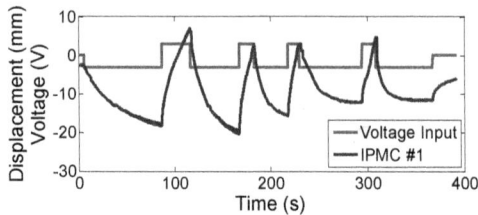

(a)

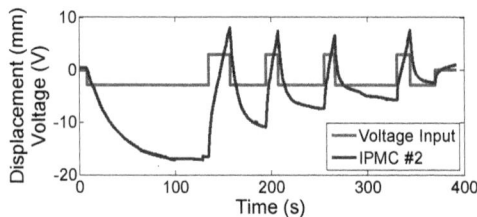

(b)

Fig. 4.4. Laser displacement of the IPMCs driving the stepper motor [156].

There is a reasonable correspondence between the simulated motor and the actual experimental results. It can be seen that at the beginning the IPMC has a larger displacement and it starts to degrade in performance. This is mainly due to the fact that the motor is operating in air for a period of time and the IPMCs exhibit highly time-varying behavior in this type of environment as the IPMC can rapidly dehydrate. It is expected that a more stable performance may be achieved if operated in water. Despite this the motor does act as the simulation predicts and rotary motion is achieved. The camera shots in Fig. 4.5 show the motor in operation, and it can be seen that the system does operate as predicted to achieve full 360° motion.

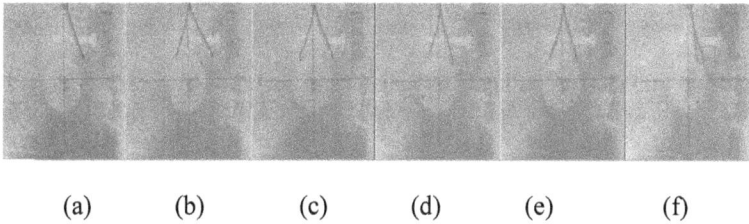

(a)   (b)   (c)   (d)   (e)   (f)

Fig. 4.5. Camera shots of motor in operation (a) at initial position (b) after step 1 (c) after step 3 (d) after step 5 (e) after step 7 (f) 315° rotation completed and IPMCs back to home position (Note: the black dot) [156].

4.4. Extension to Many IPMCs

It has been demonstrated that the stepper motor works as simulated and therefore it is valid to believe that this model and simulation technique can be extended to other devices. The next step to improving the performance of the stepper motor is to remove the pause in the operation to achieve continuous rotation as with traditional rotary motors.

A new design incorporating 4 pieces of IPMCs and using the same principle as the previous design has been proposed. The IPMCs work in two pairs, with each pair acting similarly to the two IPMCs in the previous design. In simulation it is shown that the two pairs must be 90° or more out of phase from each other to avoid the IPMC clashing when returning to the home position. The phase also has to be a multiple of 45° to remove the pause and achieve constant motion. It has therefore been designed that the IPMC pairs are 135° out of phase as seen in Fig. 4.6. The simulation results are shown in Fig. 4.7 and prove that this removes the pause in the system and that the new design can indeed achieve continuous motion.

Fig. 4.6. Motor shaft and IPMC setup for four IPMC stepper motor.

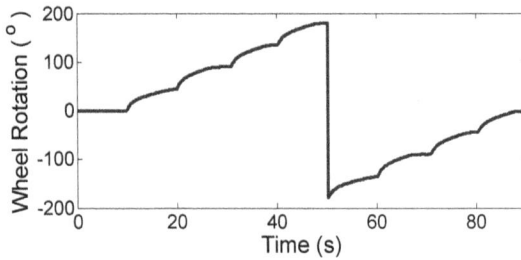

Fig. 4.7. Motor shaft angular displacement
for a four IPMC stepper motor [156].

With the four IPMC motor design there will still only be one IPMC in contact with a pin at any point in time and so, assuming the same IPMCs are to be used as in the two IPMC design, the compliance and holding torque will be the same i.e. 720 rad/Nm and 1.212×10^{-4} Nm respectively.

The time step can be altered to speed up the motor, but there is a limit to the step time as the IPMCs still need to have enough time to reach their home positions. A smaller voltage could also be applied, but this would decrease the available torque output of the motor and will decrease the speed of the motor as well.

By adding more IPMCs the step size will decrease and the motor motion will become much smoother. As the step size and hence IPMC deflections are reduced, the time to return to the home position will reduce and the speed of the motor can be increased. With more IPMCs the motor can operate with the IPMCs actuating in a wave type mode where each IPMC advances the shaft a small distance, then the next IPMC advances it another step by pushing the same pin. In this mode only one IPMC would be touching a pin at any point in time so the compliance and holding torque would be the same as previously calculated. Alternatively there could be more pins on the wheel and then more than one IPMC would be in contact with a pin at a time. This will increase the torque output of the motor but consequently the stiffness will increase reducing the compliance of the device. All these variations on the design can easily be simulated using the model to verify their design before going on to build the system. A 12-IPMC actuated stepper motor (6 each layer) is shown in Fig. 4.8.

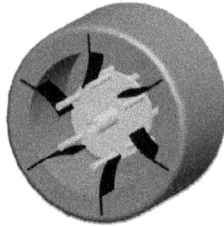

Fig. 4.8. Proposed 12-IPMC stepper motor and casing.

4.5. Discussion

This chapter describes the development of a completely novel and innovative design for a traditional device using non-traditional actuators. This is a completely new way of thinking and designing systems using smart materials. The stepper motor has been designed and implemented with two IPMCs and can successfully achieve full 360° rotation, but has some pause time in between steps. An improved design has been proposed and validated in simulation which also can achieve full rotation and removes the pause time, resulting in continuous rotation similar to a traditional stepper motor.

The IPMC model has been shown to perform for this design of a stepper motor, but more importantly than this particular design is the demonstration that this modeling and design process works well for developing real world devices.

From the experimental results in Fig. 4.4 it is clear that without close-loop control the IPMC performance can change unpredictably over a period of time. In this application for the stepper motor it is acceptable because as long as the IPMC pushes the pin, the step will occur and the system will operate as required. In many other applications a much more precise control of the IPMC displacement will be required. This has highlighted the need for designing closed-loop controllers for IPMCs in most applications. Therefore, this is the approach that will be taken for developing the remaining devices.

Chapter 5

Iterative Feedback Tuning: Fundamental Theory and Application to IpMCs

In order to harness the wide-ranging advantages of IPMC actuators and to aid their successful implementation into real systems, the actuation response of an IPMC must be effectively controlled. However, this is not a trivial task due to their complex behavior. After analyzing the experimental open-loop results of the IPMC stepper motor in the previous chapter it is clear that for precise IPMC control closed-loop architecture will be required.

IPMCs need accurate and robust controllers; the problem is these controllers typically require accurate and complete dynamic models of the IPMC response. A major problem arises as such models are not readily available mainly as the time-varying and stochastic IPMC response renders models inaccurate over long operating times. The model developed in Chapter 3 falls into this category as it is accurate at the time of modeling but may not remain so as the IPMC dynamics drift after some operating duration. This model is proficient for mechanical design, but like most IPMC models is time invariant and so is not capable for use as an accurate reference model for control.

Currently most IPMC closed-loop controllers are tuned or adapted using an approximate model. The major issue with this is that the IPMC behavior varies considerably through operation and the system dynamics will drift far from any derived model. This will cause the actual performance to vary far from the simulated or expected performance and hence there are no definite guarantees on the system reliability.

Iterative feedback tuning is an iterative tuning method which adapts the controller parameters to optimize the system performance based solely on experimental information from the actual system. As the controller is completely derived from experimental data, no model or knowledge of the system is required. Iterative feedback tuning therefore has the ability to overcome the major issue associated with controlling IPMCs caused by the tendency of the IPMC dynamics to drift far and unpredictably throughout operation. Adaptations on the standard iterative feedback tuning algorithm are developed in the next three chapters to couple the advantages of the standard algorithm with improvements in order to tackle some of the other issues with IPMC control and specific application requirements, like nonlinearities and robustness.

5.1. Iterative Feedback Tuning Background

The interesting idea of using the system itself, as opposed to a system model, to adapt the controller parameters was first proposed in 1964 by Narendra and Streeter [158]. It wasn't until 30 years later this idea reappeared in the form of the iterative feedback tuning algorithm. Iterative feedback tuning, originally proposed in 1994 by Hjalmarsson *et al.* [159], is an iterative optimization approach to designing controllers through the objective of minimizing a controller design criterion of an unknown plant. This relatively new tuning method tests the response of the actual system to determine new updated and improved control parameters. As the updated parameters are based on experiments on the actual system, the approach is model free.

Iterative feedback tuning has been implemented demonstrating good results in both laboratory and industrial applications such as control of profile cutting machines [161], speed and position control of a servo drive [161], temperature regulation in a distillation column [162], a DC servo with backlash [162], control of photo resistant film thickness [163], an inverted pendulum [164], magnetic suspension system [165] and active sound vibration control [166]. The major advantages over other tuning methods include; automatic tuning and therefore no need for experienced operator; model free so no knowledge of the system required; can be used for adaptive tuning; it can be implemented as an online tuning method; it can be used to tune many types of controllers and different design criterion can be used to optimize the operating performance depending on the application requirements.

Current research in iterative feedback tuning has been into new ways for adapting the controller parameters, new optimization algorithms to speed up convergence and avoid overstepping the optimal state [161].

Other research has been exploring guaranteeing the global optimal solution is found and not only a local minimum. The work presented here does not concentrate on these areas, but instead presents new variations to improve the iterative feedback tunings performance when tuning the nonlinear time-varying IPMCs which are integrated in novel biomedical robotics devices.

5.2. Motivation for Iterative Feedback Tuning with IPMCs

Currently most IPMC controllers are tuned in simulation using an approximate plant model. The performance is then assessed, also in simulation, before implementing the controller on the real system. One major issue with this method is the development of a suitable IPMC model which is complex and time consuming as the IPMC is extremely non-linear, time-variant and environmentally sensitive. No model exists which takes into account the full time-varying behavior of an IPMC which occurs as a result of dehydration and redistribution of ions in the polymer. Solely implementing a time-invariant model based control system is insufficient when actuating for a period of time.

If a model based approach is taken, the controller must be further fine-tuned on the real system to account for variability between the model and the real plant and the controller is then sample specific so it cannot be used on a different IPMC sample. Traditionally this tuning would require an experienced operator with knowledge of the system relying heavily on intuition. The developed controller would not then be directly transferable to other IPMC samples, so each individual sample in a system must be manually tuned. Consequently it is highly desirable to develop an automatic adaptive tuning method. The development of an iterative feedback tuning routine will allow the IPMC controller to be automatically tuned without the need for any model or knowledge of the system.

All current adaptive and robust controllers which have been experimentally implemented on IPMCs rely on a model based approach. This approach requires that a fixed controller and model structure (e.g. number of poles and zeros in the system transfer function) be proposed before operation begins. The controller parameters are then adapted so the actual system performance replicates the model reference. The problem with this is, as the IPMC dynamics drift far from the initial state the controller parameters may not be able to be accurately adapted to force the real system characteristics to replicate the model, as the fixed controller structure may not be sufficient to handle this. Also a model structure must be proposed for all operating conditions because if for example the IPMC is driving a load with considerably different impedance, the entire system characteristics will vary immensely.

All of the existing adaptive controllers have been developed in an ideal environment with no loads, friction etc. If the controllers are then used with the IPMC in the real world with these effects added, the dynamics of the system will change significantly and as such models become useless over large operating ranges, e.g. loaded then unloaded. The development of an iterative feedback tuning routine will allow the IPMC controller to be automatically adaptively tuned without the need for any model or knowledge of the system. As experiments on the system itself are used to calculate the updated controller parameters the system is impervious to variations in the structure of a system model. The controller is simply tuned using the actual system dynamics at that point in time.

The original iterative feedback tuning algorithm possesses all the characteristics which are necessary for adaptive tuning of the highly unpredictable and time-varying IPMC and as such serves as a good foundation for the controllers which will be used for the biomedical applications in this research. New variations on the original iterative feedback tuning algorithm will be developed to handle some of the more hard to control properties of the IPMCs when actuating external devices. This new iterative feedback tuning approach taken here is a novel way of thinking for IPMCs as up until now research emphasis has been on the modeling of materials in order to be able to control their behavior. This new model free approach to controller design presents a major step forward for IPMC technology towards its wide acceptance as a viable alternative to traditional actuators.

5.3.   Formulation of the Iterative Feedback Tuning Algorithm

In this section the formulation of the standard iterative feedback tuning algorithm is described, derived from its roots in adaptive control. Comparisons with traditional adaptive control techniques, as well as discussions on the major issues with iterative feedback tuning like stability, convergence and robustness are also briefly presented. A simple discrete single-in, single-out linear time invariant (Single input, single output (SISO) LTI) system, as shown in Fig. 5.1 will be used to describe the iterative feedback tuning system characteristics.

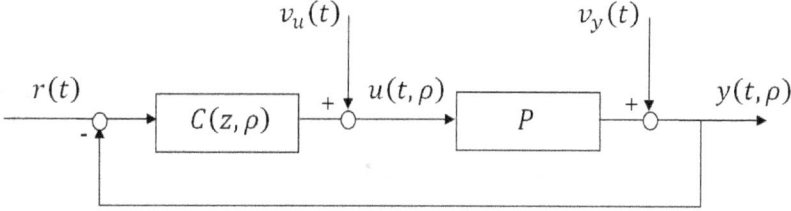

Fig. 5.1. Discrete SISO LTI close-loop control system.

Equations (5.1) and (5.2) describe the system in Fig. 5.1

$$y(t, \rho) = Pu(t, \rho) + v_y(t) \qquad (5.1)$$

$$u(t, \rho) = C(z, \rho)(r(t) - y(t, \rho)) + v_u(t) \qquad (5.2)$$

P is the unknown SISO LTI plant (z is the shift operator) and $C(z, \rho)$ is the discrete LTI controller which is a function of a vector of controller parameters $\rho \in R^{n_\rho}$. With an external deterministic reference signal, $r(t) \in R$, the signal $u(t, \rho) \in R$ is the process input and $y(t, \rho) \in R$ is the corresponding process output. $v(t) = \left[v_y(t) v_u(t) \right] \in R^2$ is an immeasurable stochastic disturbance. *r(t)* is independent of $\{v(t)\}$. In the following formulations the time, *t*, and shift operator, z, arguments will be omitted whenever not needed and all signals obtained from the closed-loop system with controller, $C(z, \rho)$, will be indicated using the ρ argument. Also the parameterization of the controller must be such that all signals in the system are differentiable with respect to. ρ.

The closed-loop response and sensitivity function of the controller C(ρ) are T(ρ) and S(ρ) respectively and is shown in equations (5.3) and (5.4).

$$T(\rho) = \frac{PC(\rho)}{1 + PC(\rho)} \qquad (5.3)$$

83

$$S(\rho) = \frac{1}{1 + PC(\rho)} \qquad (5.4)$$

5.3.1. Minimizing a Performance Criteria

The objective of a controller design is to minimize some objective function in order to optimize the system response. If we let y_d be the desired trajectory, then the error between the desired and achieved response is

$$\tilde{y}(\rho) = y(\rho) - y_d \qquad (5.5)$$

It then makes sense to formulate the function to be minimized or the design criteria based on this output tracking error, $\tilde{y}(\rho)$. Using a quadratic function based on a least squares fit gives the design criteria, $J(\rho)$, equal to

$$J(\rho) = \frac{1}{2N} \sum_{t=1}^{N} E\left[\tilde{y}(t, \rho)^2 \right], \qquad (5.6)$$

where $E[\ ]$ denotes the expectation with respect to the disturbance v and N is the number of samples in an experiment. The iterative feedback tuning algorithm discussed here is valid for almost any signal based objective function [167]. A number of variations have been proposed in various literatures, frequency and time weighted filters can be added to put more emphasis on one or another aspects of the system, for example overshoot, or steady-state response. The control effort $u(t, \rho)$ is commonly used with the output error in the design criteria; this prevents tuning towards excessively high gains and hence causing saturated inputs into the system as well as preventing high overall power consumption. For the IPMC systems in this research the design criteria will not include the control effort as it has been shown that IPMCs do have inherently low power consumption and therefore the most important factor in controller and device design is the tracking error. As such the design criteria in equation (5.6) will be used for all IPMC experiments in this research.

The necessary condition for optimal performance of the system is that the first derivative of the design criteria, with respect to the controller parameters, ρ, is equal to zero.

$$\frac{\partial J}{\partial \rho}(\rho) = \frac{1}{N}\sum_{t=1}^{N}E\left[\tilde{y}(t,\rho)\frac{\partial \tilde{y}}{\partial \rho}(t,\rho)\right]$$

$$= \frac{1}{N}\sum_{t=1}^{N}\left[\tilde{y}(t,\rho)\frac{\partial y}{\partial \rho}(t,\rho)\right] = 0 \tag{5.7}$$

To find the optimal performance for the system with any controller, this condition must be able to be detected by finding the following quantities

(i)   the signal, $\tilde{y}(t,\rho)$

(ii)   the gradient of the output, $\dfrac{\partial y}{\partial \rho}(t,\rho)$

The signal $\tilde{y}(t,\rho)$ can easily be calculated by measuring the output of the system with a given reference and then taking that desired reference away from the output, as in equation (5.5). It is the gradient signal which is problematic to calculate. The reason for this can be seen through the following explanation. The differentials of equations (5.1) and (5.2) with respect to ρ are,

$$y'(t,\rho) = Pu'(t,\rho) \tag{5.8}$$

$$u'(t,\rho) = C'(z,\rho)(r(t) - y(t,\rho)) - C(z,\rho)y'(t,\rho) \tag{5.9}$$

The full derivative can then be expressed as

$$\frac{\partial y}{\partial \rho}(t,\rho) = PS(z,\rho)\frac{\partial C}{\partial \rho}(z,\rho)(r(t) - y(t,\rho)), \tag{5.10}$$

where $\dfrac{\partial y}{\partial \rho}$ represents the full derivative with respect to the parameter vector ρ. In equation (5.10), $\dfrac{\partial C}{\partial \rho}(z,\rho)$ is a known function of ρ, which depends on the parameterization of the controller $C(z,\rho)$, while the sensitivity, $S(z,\rho)$, depends on the unknown system and therefore some information from the system is required. This is where the problem of adaptive control occurs.

Many of the early adaptive controllers were developed using a direct approximation of the gradient of the design criteria, $\frac{\partial J}{\partial \rho}$ itself, with a gradient search and minimization similar to equation (5.11), in order to calculate a parameter update.

$$\rho(i+1) = \rho(i) - \gamma R^{-1}(i)\frac{\widehat{\partial J}}{\partial \rho}(\rho(i)), \qquad (5.11)$$

where i is the iteration number, γ is a positive real scalar to determine the step size, $R(i)$ is some positive definite matrix to determine the search direction and $\frac{\widehat{\partial J}}{\partial \rho}$ is an approximation of the gradient.

Over the years a number of other methods to find approximations for $\frac{\partial J}{\partial \rho}$ have been proposed. This is the basis for traditional adaptive control, to which iterative feedback tuning is built on.

One of the first ideas to obtain $\frac{\widehat{\partial J}}{\partial \rho}$ was to use a parameter perturbation method which uses a numerical approximation approach. This approximation is found by evaluating a sample version of the design criteria, $J(\rho)$ in equation (5.6), for close perturbed values of ρ and then using these to approximate a gradient $\frac{\partial J}{\partial \rho}$, see Refs [168, 169]. The main problem with this method is that the number of experiments required is proportional to the number of parameters and thus this method is very slow. New methods to reduce the number of experiments were developed in Ref [170].

Sensitivity model methods were later developed when it was realized that for a LTI process the sensitivity of any signal in the system with respect to any scalar parameter, e.g. is ρ, could be computed using a LTI differential equation. This meant that for any signal based design criteria the parameter sensitivities could be found using a simulation based on a model of the process and then the gradient based minimization in equation (5.11) can be employed [171,172].

Model reference adaptive control (MRAC) was introduced in [173] when a way around having to develop a model of the plant was devised. It was observed that the purpose of a control system is to tune towards some desired close-loop performance such that $T(\rho) \approx T_d$ (where T_d is the desired close-loop response) so then if one assumes this already holds then $S(z, \rho)$ in equation (5.10) can be replaced with $1 - T_d$. Using this reference model approach with the gradient based pure steepest-descent search became famously known as the M.I.T rule. Another approach is a self-tuning controller [174], which uses recursively identified models of the system for calculating the parameter update.

The problem with the self-tuning and MRAC as well as many other versions of these control methods developed more recently is they require fixed model structures which may not be suitable for a system with large varying dynamics and loads as the IPMC systems in this research. As such a different approach from traditional adaptive control was sought for this research and IFT has been identified as the adaptive control system which has the most promise for the application due to its independence from a system or reference model. IFT is robust to the underlying properties of the system compared with the model based methods [167].

5.3.2. Standard Iterative Feedback Tuning algorithm

In [159] it was proposed that the real closed-loop system could be used to calculate y' and u' in equations (5.8) and (5.9) and hence $\dfrac{\partial y}{\partial \rho}(t, \rho)$ in equation (5.10). This is done using the following basic IFT algorithm:

1. Perform a 'normal' experiment on the closed-loop system, Fig. 5.1 and equations (5.1) – (5.2), using a standard reference, $r(t)$. Measure N samples of the system output which can be expressed as,

$$y(\rho) = T(\rho)r + PS(\rho)v_u + S(\rho)v_y \qquad (5.12)$$

This first experiment is referred to as the normal experiment, as it is performed in normal experimental conditions. The superscript 1 will be used to denote signals from this experiment, i.e. r^l and y^l are the reference and output from this experiment, respectively.

2. Inject the signal error, $\tilde{y}(\rho)$, from the first experiment as the new reference for the second experiment, i.e. $r - y^l$. This second experiment is called the gradient experiment and signals from this experiment will be denoted with the superscript 2. The output from this experiment is then

$$y^2(\rho) = T(\rho)(r - y^1) + S(\rho)(Pv_u^2 + v_y^2) \qquad (5.13)$$

3. Take the gradient approximation of the system output as

$$\frac{\widehat{\partial y}}{\partial \rho}(\rho) = \frac{1}{C}(\rho)\frac{\partial C}{\partial \rho}(\rho)y^2(\rho) \qquad (5.14)$$

This is found by passing y^2 through the transfer function, $\frac{1}{C}(\rho)\frac{\partial C}{\partial \rho}(\rho)$, as per Fig. 5.2 where the final output $\widehat{y}$ is the derivative with respect to one element in ρ. For implementation of a real system $\dfrac{C'(\rho)}{C(\rho)}$ is a vector of transfer function where each element corresponds to the derivative of the controller with respect to one of the controller parameters divided by the controller.

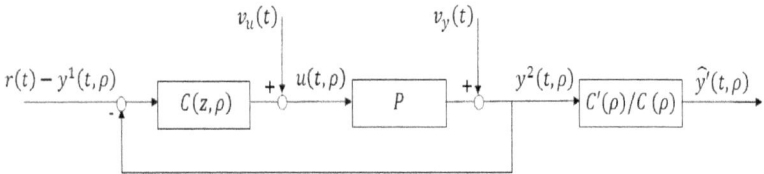

Fig. 5.2. Gradient experiment.

The estimate of the gradient using IFT can be expressed as equation (5.15) by combining equations (5.13) and (5.14) then comparing this with equation (5.10).

$$\frac{\widehat{\partial y}}{\partial \rho}(\rho) = \frac{\partial C}{\partial \rho}(\rho)\left(PS(\rho)(r - y(\rho)) + \frac{1}{C}(\rho)S(\rho)(Pv_u^2 + v_y^2) \right)$$

$$\frac{\widehat{\partial y}}{\partial \rho}(\rho) = \frac{\partial y}{\partial \rho}(\rho) + \frac{1}{C}(\rho)\frac{\partial C}{\partial \rho}(\rho)S(\rho)\left(Pv_u^2 + v_y^2 \right) \qquad (5.15)$$

$$= \frac{\partial y}{\partial \rho}(\rho) + w(\rho)$$

When comparing with the actual gradient, it can be seen that there is a perturbation, $w(\rho)$, introduced which causes a bias in the gradient approximation due to the disturbances, v_u^2 and v_y^2 in the second experiment as seen in equation (5.16)

$$w(\rho) = \frac{1}{C}(\rho)\frac{\partial C}{\partial \rho}(\rho)S(\rho)(Pv_u^2 + v_y^2) \qquad (5.16)$$

The advantage of this IFT algorithm, as explained, is that it includes no approximation of $T(\rho)$, the closed-loop system transfer function, however due to the disturbances in the second experiment there may be some error in the approximation of $\frac{\partial J}{\partial \rho}$. Under mild assumptions it can be shown that this error adds a very small contribution to the estimate, as explained in the next section.

An estimate of the gradient of the design criteria, equation (5.17) is found using $\tilde{y}(t, \rho)$ and approximation of the gradient, $\frac{\widehat{\partial y}}{\partial \rho}(\rho)$, found through the steps 1-3, above, and this is then used with the iterative method in equation (5.11) to optimize the system response.

$$\frac{\widehat{\partial J}}{\partial \rho}(\rho) = \frac{1}{N}\sum_{t=1}^{N} \tilde{y}(t, \rho)\frac{\widehat{\partial y}}{\partial \rho}(t, \rho) \qquad (5.17)$$

It can be seen that the IFT does indeed have its roots in standard adaptive control, yet through utilizing the actual system instead of an approximate model of the system to determine the gradient estimate, IFT is more robust to the drift in system dynamics and simpler to implement as there is no need to obtain a system model. Also the error in the IFT approximation for $\dfrac{\partial J}{\partial \rho}$ is only due to the disturbances in the second experiment, which as explained next are very small, while the error in the approximation for $\dfrac{\partial J}{\partial \rho}$ for model-based methods is due to the difference between the actual system and the system model used. This is likely to be much larger than that of $w(\rho)$ for IFT.

5.3.3. An Unbiased Gradient Estimate

The perturbation $w(\rho)$ introduces a bias error due to the disturbances $v = \begin{bmatrix} v_y & v_u \end{bmatrix}$. If these disturbances are considered under the mild assumption that they are stochastic variables with zero mean and that the disturbances v^1 and v^2, from the normal and gradient experiments, respectively, are uncorrelated, then the gradient approximation in equation (5.17) is unbiased despite the perturbation $w(\rho)$. This is validated from:

$$
\begin{aligned}
E\left[\widehat{\frac{\partial J}{\partial \rho}}(\rho) \right] &= \frac{1}{N}\sum_{t=1}^{N} E\left[\tilde{y}(t,\rho)\widehat{\frac{\partial y}{\partial \rho}}(t,\rho) \right] \\
&= \frac{1}{N}\sum_{t=1}^{N} E\left[\tilde{y}(t,\rho)\frac{\partial y}{\partial \rho}(t,\rho) \right] + E[\tilde{y}(t,\rho)w(t,\rho)] \\
&= \frac{\partial J}{\partial \rho}(\rho) + \frac{1}{N}\sum_{t=1}^{N} E\left[\tilde{y}(t,\rho) \right]E[(w(t,\rho)] \\
&= \frac{\partial J}{\partial \rho}(\rho) + 0
\end{aligned}
\qquad (5.18)
$$

where in the last step the zero mean assumption of v has been used. The unbiased property is the key characteristic of IFT. This means an accurate approximation for the gradient can be obtained, also this is very important to prove convergence of the design criteria.

5.3.4. Convergence, Stability and Robustness

Using classical results in stochastic approximation the convergence of the design criteria to a stationary point can be proved regardless of the order of the underlying LTI system, P, and the complexity of the controller, C [167]. The basic requirement for this convergence is that all signals remain bounded throughout the iterations. A complete derivation of the convergence result can be found in [167]. The power of the theorem presented is that there are no assumptions on the system and controller except that they are LTI, thus the results apply to a simple PID controller as well as to a much more complex system.

In [159] it is briefly discussed that provided the initial control system is stable, if the step size is small enough and the data set is relatively large then the search will always be in the negative direction, ensuring convergence to the local minimum of the design criteria.

It is important to have a robust closed-loop design in order to ensure the stability of the system. For the model-based approaches this is achieved by choosing robustness margins which are large enough to cope with the model uncertainty and drift in system dynamics. A problem does occur however if the system leaves these margins. With IFT two major aspects must be considered to ensure the robustness and hence stability, firstly the objective function must be chosen wisely so the optimal controller does not correspond to one with poor stability margins. Secondly the iterative minimization itself can cause problems. Even if for example the exact gradient is known, the update may lead to an unstable controller if the step size is chosen too large and the parameters overshoot the optimal value. It is proposed in [160] to use a varying step size in order to reduce the step size as the optimal approaches. Also in [160] and [103] it is proposed that the design criteria can be evaluated at each step using the normal experiment, and if this is not smaller than the previous design criteria, then the controller must have 'overstepped' the optimum and thus the step size is reduced and no parameter update is made until the step size is sufficiently small enough to prevent this overshoot.

In many systems some knowledge of the process is likely to be known. For example certain parameters may have more or less effect on the system dynamics, and certain parameters may be known to be closer to the optimum than others. For this reason different step sizes can be used for different controller parameters to adjust their update rate. This can also improve stability if for example a parameter is likely to cause the system to go unstable if it is increased too high, for example a derivative gain can amplify system noise causing instability by invoking some high un-modeled resonance.

It is also useful to include a penalty on the control effort to restrict the controller parameters from rapidly increasing leading the system to saturate and even causing unstable response. This is also useful for limiting power consumption. The design criterion then resembles that of a LQR type controller. Design criteria which filter certain frequencies, penalizing operation around some unwanted resonances which may cause instability, can be implemented making the system more robust. Also time weighting filters, which may put large penalties on overshoot or oscillation, may be included to stabilize the system response.

5.3.5. Search Direction

The matrix $R(i)$ in equation (5.11) determines the update direction for the controller parameters and is therefore a very important design choice in IFT. Using the identity matrix for $R(i)$ gives a negative gradient direction. It is commonly accepted in literature [27, 29, 31] that using the Gauss-Newton approximation of the Hessian for $R(i)$ gives improved results, this becomes more important when the sample size is small. The Hessian is given below in equation (5.19).

$$R(i) = \frac{1}{N} \sum_{t=1}^{N} \left(\frac{\partial \widehat{y}}{\partial \rho}(\rho(i)) \frac{\partial \widehat{y}}{\partial \rho}^{T}(\rho(i)) \right)$$ (5.19)

This approximation has a positive bias term, but under the mild assumptions this does not affect the convergence of the design criteria [162, 167].

5.3.6. Update Rate

Since their advent the vast majority of adaptive control systems have been developed where the parameter updates occur on a very short time scale, usually at each new sample time. This naturally results in fast tuning times but consequently makes the system nonlinear and time-varying which significantly complicates the analysis [175].

IFT and other iterative identification and control schemes, which have been developed more recently, update parameters on a much slower time scale. These methods collect data through experiments and the batches of data are processed with a fixed controller in the loop through the entire experiment. This drastically simplifies analysis, even permitting simple LTI assessment, but does result in a slower convergence rate.

5.3.7. Issue with the Gradient Experiment

The gradient approximation $\dfrac{\widehat{\partial y}}{\partial \rho}$ is found using a special gradient experiment which traditionally uses the reference, $r^2 = r - y^1$. As the error from the first experiment is used this will cause the output from the gradient experiment to vary far from the normal experiment. This places some restriction on the use of IFT in certain practical applications where the trajectory of the system should not vary far from a desired reference since this may result in production waste. Some methods have been proposed to overcome this in [167], yet very little work has been carried out to realize implementations to overcome this issue. In this research the need for a gradient experiment which varies far from the normal experiment is tackled and a new IFT algorithm is developed and implemented to allow online system tuning without production waste.

5.3.8. Extensions to Nonlinear, Time-varying and Multivariable Systems

A rationale which shows that IFT can handle non-linear processes equally well as LTI processes is given in [167] and an example of a nonlinear drive with backlash which has good IFT performance is presented in [162].

As IFT is an adaptive tuning method it is fair to assume that it can accurately handle slowly time-varying systems. Also as IFT does not assume a fixed model structure it can be used when the system dynamics change far from their initial state. IFT is more robust than model-based methods to handle time-varying systems.

IFT can be extended to multi-variable systems very easily as described in [167]. The algorithm works in much the same way but additional experiments are required depending on the number of input variables and the number of sensed outputs which are feedback.

93

5.4. Iterative Feedback Tuning Implementation on an IPMC

The standard IFT algorithm which has been described in the previous section is now implemented onto an IPMC in simple cantilever configuration as in Fig. 5.3 to verify that the IFT algorithm does work for the system. This is in a lab environment with no external disturbances as has been done previously in most research into IPMC control.

Fig. 5.3. IPMC clamped in cantilever configuration.

5.4.1. Experimental Setup

The experiments were undertaken using a custom test rig as shown in Fig. 5.3. The rig supports 2 copper clamps which act as electrodes to pass the voltage to the IPMC. A Nafion® based IPMC was used, with platinum electrodes. The IPMC was 35 mm long, 10 mm wide with a thickness of 200 μm. The clamped length was 5 mm.

The IPMC and clamps are placed in a container of de-ionized water, seen in Fig. 5.4, in order to avoid rapid dehydration and potential damage to the IPMC. This should also stop the IPMC dynamics from drifting quickly, this will allow the IFT algorithm to be assessed more independently from the time-varying characteristics of the IPMC itself. Control electronics and a National Instruments DAQ card are used to interface between the MatLab environment running the Simulink model on the PC and the IPMC actuator.

Fig. 5.4. Test rig used for the IPMC experiments.

A Banner LG10A65PU laser sensor, with the resolution set to 10 μm, was used to measure the displacement of the IPMC. It was setup as per Fig. 3.11 to measure the linear displacement at a distance of 25 mm from the base, which corresponds to 5 mm back from the tip. The laser is placed this far back from the tip to ensure that at high IPMC displacements/curvature the laser target will not leave the tip of the IPMC.

5.4.2. IPMC Control System

The control architecture which will be used to tune the IPMC will consist of proportional, integral and derivative elements. The IFT algorithm is then used to tune the parameter vector, $\rho = \begin{bmatrix} K_p K_i K_d \end{bmatrix}^T$. Fig. 5.5 depicts the control system that was developed and implemented in Simulink® for tuning the control parameters.

The switch in the system is used to control what signal is injected as the reference input. To tune for a specific displacement, a step input of that displacement is set as the reference trajectory, for the first experiment. For the second tuning or gradient experiment the switch is turned and the system has the reference input, $r^2 = \left(r - y^1 \right)$, which is the error of the first experiment. After these two experiments, the control parameters are updated and then the next iteration is run.

The discrete transfer functions expressed in terms of the shift operator, z, which describe the block diagram in Fig. 5.5 are specified in equations (5.20) to (5.23). T is the discrete sampling time for the controller which has been chosen as 0.1s for this system as the IPMC has slow dynamics and therefore 10 Hz is fast enough to accurately capture the changes in output. L_P is the low pass filter constant, set to 0.85, this is used to suppress any unwanted high frequency inputs to the IPMC caused by noise or other disturbances. The saturation block was placed in the loop and set to ±3 V to ensure that no excessive voltage was input to the IPMC preventing any damage. This also ensures that the current draw is relatively small and hence the power consumption of the IPMCs is low.

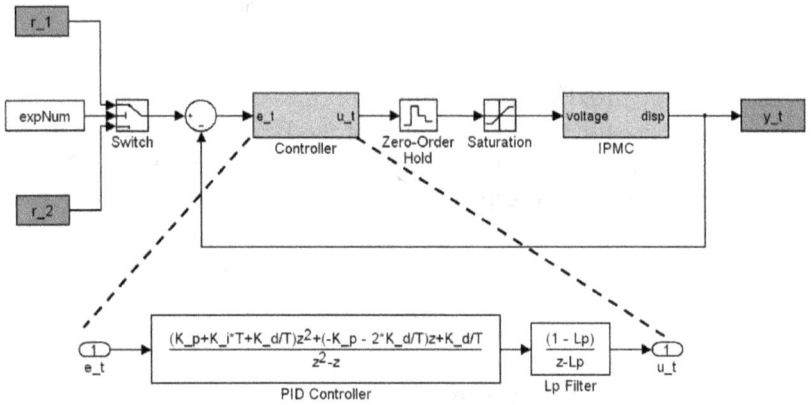

Fig. 5.5. Schematic diagram of the IPMC control system used to tune controllers.

$$PID(z) = \frac{(K_p + K_i T + \frac{K_d}{T})z^2 - (K_p + \frac{2K_d}{T})z + \frac{K_d}{T}}{z^2 - z}$$

(5.20)

$$LPF(z) = \frac{1 - L_p}{z - L_p}$$

(5.21)

$$ZOH(z) = \frac{2z}{T(z + e^{-2})}$$

(5.22)

$$C(z) = PID(z).LPF(z).ZOH(z)$$

(5.23)

The terms $\dfrac{1}{C(\rho)}\dfrac{\partial C}{\partial \rho}(\rho)$ for the IPMC control system, which are needed to solve equation (5.14) are given in equations (5.24) to (5.26).

$$\frac{1}{C(z)}\frac{\partial C(z)}{\partial K_p} = \frac{z^2 - z}{\left(K_p + K_iT + \dfrac{K_d}{T}\right)z^2 - \left(K_p + 2\dfrac{K_d}{T}\right)z + \dfrac{K_d}{T}} \tag{5.24}$$

$$\frac{1}{C(z)}\frac{\partial C(z)}{\partial K_i} = \frac{Tz^2}{\left(K_p + K_iT + \dfrac{K_d}{T}\right)z^2 - \left(K_p + 2\dfrac{K_d}{T}\right)z + \dfrac{K_d}{T}} \tag{5.25}$$

$$\frac{1}{C(z)}\frac{\partial C(z)}{\partial K_d} = \frac{1}{T}\frac{z^2 - 2z + 1}{\left(K_p + K_iT + \dfrac{K_d}{T}\right)z^2 - \left(K_p + 2\dfrac{K_d}{T}\right)z + \dfrac{K_d}{T}} \tag{5.26}$$

5.4.3. Iterative Feedback Tuning Settings

The design criteria in equation (5.6) will be used as the design criteria for the IPMC IFT experiments. As stated previously the control effort, $u(t,\rho)$, is commonly used with the output error in the design criteria to prevent excessively high gains causing saturation as well as reducing overall power consumption. It is well documented that IPMCs have inherently small power consumption and hence the most important factor in design is the tracking error and so the design criteria in equation (5.6) is the most appropriate. When IPMC systems are used in real embedded applications where the power supply is a major issue, a penalty on high control effort should be included.

Due to the desired applications in biomedical robotics, the control system was designed to be accurate for changes in set point, in terms of both the transient and steady state response. In order to tune for this, the reference trajectory was a stair-step function to a desired set point to be tuned. The experiments will be 60 s long consisting of, +3 mm for the first 15 s, then step back to zero displacement until 30 s, then -3 V until 45 s and finally to zero displacement until 60 s. This reference will ensure that the IPMC has been tuned in both directions, as it has been shown that due to imperfect fabrication techniques the IPMC can have different performance in different directions, and will also tune for 4 transient periods as well as steady state behavior.

The initial controller parameters were chosen by simulating the IPMC, with the model developed in Chapter 3, to find controller parameters which will give a stable system but with sub optimal performance. K_P was set to ensure that at the first time step, the error for the largest desired displacement would not saturate the control output. The integral gain was chosen conservatively so it does not introduce too much oscillation but still ensures zero steady state error. The derivative gain in a PID controller contributes based on the change in error, and therefore will amplify any high frequency noise that may be present in the laser sensor or control electronics. It was desired to control the IPMC to micrometer displacements, where the noise starts to become an appreciable part of the feedback signal, so a high K_d value is likely to introduce large high frequency oscillation and possibly make the system unstable. Also it has been shown by Liu in 2010 [35] that PI controllers can exhibit good response in controlling IPMCs. For these reasons it was decided to start with a PI controller by setting the derivative gain to zero and let the tuning algorithm decide how much derivative action to include. The chosen initial values were, $\rho = [1000 \; 500 \; 0]^T$.

In order to ensure convergence to the local minimum of the design criteria the data set has been chosen large, 600 samples (60 s at 10 Hz), and the step size for the control parameters is chosen to be relatively small. The step size must be small enough to ensure the controller does not 'jump too far' and result in an unstable system, but be large enough so that there is a rapid convergence to the minimum design criteria, otherwise too many iterations will be needed, making the algorithm impractical. The step sizes chosen for the IPMC system were $\gamma_{K_p} = 1; \gamma_{K_i} = 1$ and $\gamma_{K_d} = 0.5$. As a rule for the IPMC system the step size was chosen so that control parameters would update by no more than 100% of the previous value. From the experiments undertaken it has been shown that this step size will ensure that the system will remain stable, but also achieve a rapid convergence within 5 iterations. The value for step size of the derivative gain was chosen as half of that for the proportional and integral gain because for a large increase in derivative term it is possible the system may iterate to an unstable system at low deflections in the presence of large noise input.

5.4.4. Results

With the setup completed the IFT algorithm was run on the IPMC starting from the initial controller values and step sizes. The time response for a 3 mm target displacement is shown in Fig. 5.6. First is the initial output using the PI controller found in simulation, see Fig. 5.6 (a), then the performance of the IPMC with the next five consecutive controller parameters which are iteratively found by the IFT algorithm are shown, see Fig. 5.6 (b) – (f) respectively.

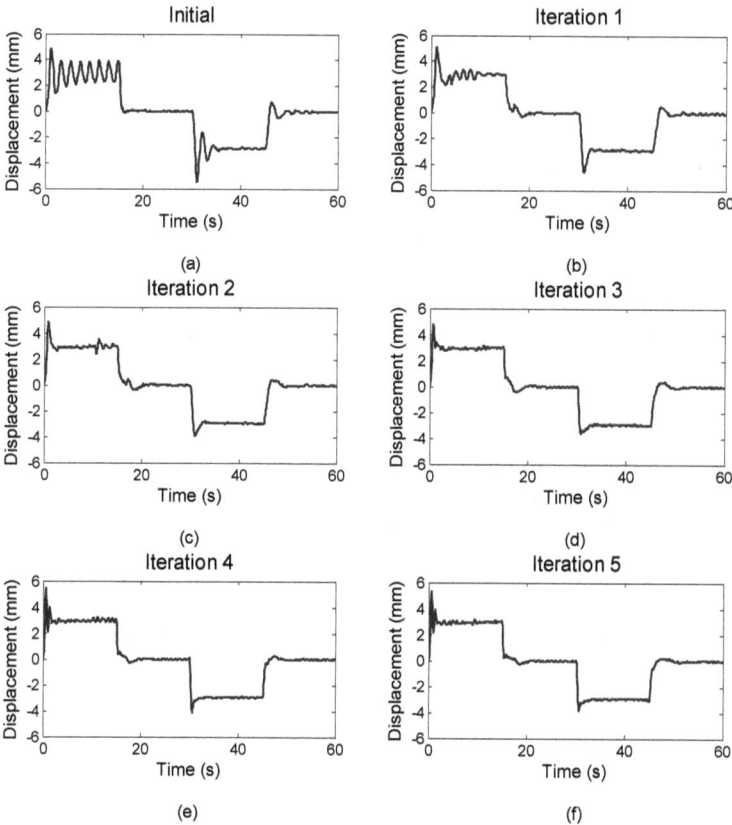

Fig. 5.6. Time response over 5 iterations of the controller for 3 mm step displacement [19].

It is clear to see that the initial controller had large oscillation in the first quarter, large overshoot starting at 30 s and also some overshoot and oscillation when finally returning to zero displacement. Even after only one iteration of the control parameters, there is an obvious improvement in the response, with the oscillations in the first 15 s reduced significantly as well as the overshoot at 30 s. After each iteration, the controller parameters are updated and it can be seen from the time response that the IPMC output drastically improved. After 5 iterations the performance of the controller for a 3 mm displacement had significantly improved. The improvement can be quantified as a 56 % improvement of the actual experimental design criteria, $J(\rho)$, equation (5.27).

$$J(\rho) = \frac{1}{2N} \sum_{t=1}^{N} \tilde{y}(t,\rho)^2 \qquad (5.27)$$

5.5. Discussion

The main motivation for employing closed-loop control is to reliably guarantee a system's performance. The actuation response of an IPMC must be effectively controlled in order to harness the wide-ranging advantages of IPMC actuators and to aid their successful implementation into real systems.

IFT has a number of characteristics which make it more desirable for use with IPMC's than other control architectures. IFT tunes the system with no requirement of a model or knowledge of the system and as such removes the time-consuming task of model development. Also as the system parameters of the IPMC will drift far from any fixed model structure which is proposed by a model-based adaptive controller, IFT will be more robust in this application. IFT has the advantage of using simple linear analysis to analyze the response due to the batch update rate. IFT can handle nonlinear, time-varying and multi-variable systems which make it useful for IPMCs and their applications.

The hysteresis behavior and other nonlinearities are not directly addressed by the IFT controllers, as some model or knowledge of the system would be necessary to account for this [34], which is specifically what the IFT algorithm is avoiding. Despite this the controller may well instinctively compensate for some part of the hysteresis as it is automatically tuning online as the system operates in certain modes.

Here, the design criteria has been chosen as a function of tracking error as this is the most important measure of the devices' performance. As IPMCs consume little power when the voltage is restricted to ±3 V no penalty has been place on the control effort. In all the IPMC experiments the controller rarely nears this saturation point. As the control effort tends to be far from the safe level set by the saturation and the control effort is not included in the design criteria, hence the control effort data for the following applications is not included as this is not relevant for measuring the device and IFT performances. If the devices are to be implemented in real remote and embedded applications where power supply is limited then the control effort should be considered.

As IPMCs are very complex and each application has its own performance criteria, it is necessary to tailor a controller's properties for each specific application. No one controller can be used for all IPMC applications. In this research the standard IFT algorithm which has been described here and implemented on the IPMC will be adapted to develop new IFT control systems to handle the specific requirements for each of the following applications.

5.6.   Iterative Feedback Tuning Summary

In this chapter a background and formulation of the standard IFT algorithm has been presented. The major issues and downfalls with the current IFT algorithm have also been discussed. The IFT algorithm has then been implemented on an IPMC in the standard cantilever configuration and has proven to be able to successfully control the tip displacement for a step target input. The response of the system starts from an initially un-tuned state with high oscillations and then the IFT has improved the performance by 56 %. This has proven the success of IFT with IPMCs and as such this research will move forward, developing new variations of the IFT algorithm for implementation with novel IPMC actuated biomedical robotics applications. The new variations will extend the current capabilities of IFT to allow implementation on new classes of systems and show the ability of IFT to control the time-varying properties of IPMCs.

Chapter 6

Robotic Rotary Finger Joint with Iterative Feedback Tuning Gain Scheduled Control

A variety of research work has been undertaken on free bending IPMC actuators, but very little work has been carried out using them to drive actual mechanisms. The research in this chapter explores the potential for applying IPMCs to drive rotary mechanisms for biomedical robotic applications.

The motivation is to develop a compliant, back-drivable robotic joint which demonstrates the potential for IPMCs driving mechanisms particularly those suited to biomedical applications such as joint prosthesis, surgical robotics, soft micromanipulators, biomimetic micro robots etc. The joint design is inspired by biological systems, with the rigid arm replicating a skeletal system and the flexible IPMC actuator, which is attached and drives the arm, acting similarly to a biological muscle actuating the limb bones. In this way the IPMC acts as an artificial muscle actuator. In the future this idea may be extended to an IPMC being implanted to replace a damaged muscle in humans suffering from multiple sclerosis or other muscular degenerative diseases.

IPMCs are very well suited to this application in human muscle augmentation mainly as they as they exhibit properties which can mimic human muscle and hence lightweight integrated biomimetic devices which can operate seamlessly with humans can be designed. Other advantages IPMCs have when compared with other actuation technologies for this application include, higher displacements than piezoelectric actuators, lower currents and higher efficiencies than smart memory alloys, also they require thermal heating which is impractical for this application. Electronic electro-active polymers, for example dielectric elastomers, require high voltages and mechanical pre-stretch mechanisms which makes them less suitable than IPMCs.

Other ionic electro-active polymers like conducting polymers typically require encapsulation throughout actuation and as such IPMCs are considered to show the most promise for this application.

Rotary joints are commonly found in robotics and industrial applications as well and so this design is also applicable to a much wider range of applications that would benefit from lightweight, flexible actuators driving mechanisms such as mechanical linkages, industrial robots, positioning systems for laser sensors, mirrors and cameras also in unmanned aerial vehicles, pick and place robots and MEMS devices, which account for a sizeable portion of modern technology. Furthermore, real life issues such as mechanism dynamics, friction and weight are tackled so these actuators could potentially be used as replacements for existing, more bulky devices.

As the system is particularly intended for biomedical robotics applications it is tuned for step changes in set point and operation at low frequencies for a large range of displacement outputs on both the micro ($<$1 mm) and macro ($>$1 mm) scale using the IFT algorithm. A schedule to vary the controller parameters, with respect to the IPMC state, is developed so the system can operate accurately and seamlessly through the large displacement range. This is essential as a finger joint must be capable of moving quickly over large displacements, yet also have the precise and sensitive control to achieve delicate tasks. This is not easily implemented with traditional motors as when they are geared for precise operation; they are very slow and stiff and so cannot achieve large displacements quickly and are also then not back-drivable.

Reference signals ranging from micro to macro range are considered for the IPMC. The developed gain scheduled (GS) nonlinear controller adapts itself in order to tackle the nonlinearities over this displacement range. No previous controllers described in literature at the time of this writing have shown they can achieve this, due to the nonlinear nature of the IPMC and the varying dynamics of the IPMC at micro and macro displacement ranges.

A first version of the rotary joint was presented at the 2009 SPIE conference on Electroactive Polymers and Devices [119], the new gain-scheduled IFT controller has been published in the Sensors and Actuators A: Physical journal [19] and the implementation of the IFT tuned control system on a new rotary joint is published in the Mechatronics journal [176].

6.1.   Mechanism Design

A simple lightweight, rigid, single DOF rotary joint has been designed which incorporates an IPMC in the commonly used cantilevered configuration, as shown in Fig. 6.1. The arm is rapid prototyped using ABS material. The rotary linkage has a length of 30 mm, a width of 20 mm wide and total weight of 1.1 g. This design can accommodate a number of different lengths of IPMC actuator. At the end of the rotary linkage is a slot compartment. The IPMC pushes against the side walls of the slot when actuated to move the rotary linkage. When assembled, the rotary linkage can be driven up to 40° in each direction which exceeds the capabilities of the IPMC actuators based on the results of the open-loop deflection experiments. Results from twenty consecutive friction tests revealed that the average blocking force of the rotary mechanism is 0.084 gf with a standard deviation of 9.725×10^{-5} gf.

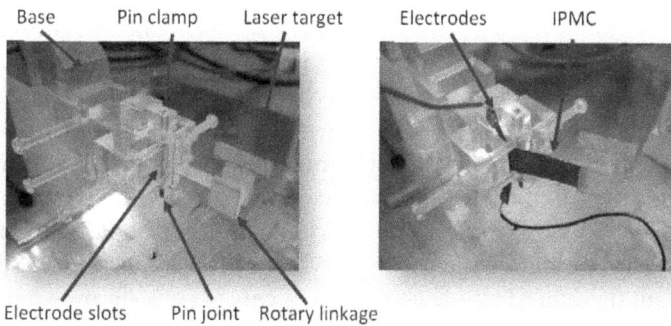

Fig. 6.1. Photo of the rotary mechanism (a) without IPMC and (b) actuated by an IPMC.

This biologically inspired design has the major advantage that the rigid arm (skeleton) will take most of the loading and pressure when interacting with the environment and the soft IPMC actuator (muscle) interacts with the environment by applying force onto the arm. In this way much larger forces and loads can be handled by the IPMC in a more robust way.

6.1.1.   Extension to Full Hand Exoskeleton/Prosthesis

Currently there is a number of hand prosthesis being developed, but all use bulky and heavy motors to move each finger joint which adds up to a significant weight and complexity of design as there are over 14 joints to move four fingers and the thumb. Also these designs can only be used for patients who have completely lost their hand and not for a patient who still has their hand, but has limited or no muscle control over its motion for example is suffering from multiple sclerosis or Parkinson's disease. It is practically impossible to achieve hand augmentation with traditional motors if the system is to be functional for someone with a muscular disease as they will not be able to handle the weight of the device. Also as motors are very rigid they can easily apply a force onto the patient which is too high causing some further pain and even injury.

Extending the research into a finger joint, a new exoskeleton as shown in Fig. 6.2, has been proposed. The system incorporates a bending IPMC actuator at each finger joint, which can all be controlled independently to augment the force of the human patient. In this way a patient will be able to grasp and lift objects they normally would not be able to. As the IPMCs are thin strips they can easily be integrated into a lightweight device similar to a glove that can easily be worn over the hand. In this way the device can be used to assist patients with muscular debilitating diseases, who still have their hands so are not suitable candidates for full a prosthesis. As the device is not bulky it can easily be lifted by the patient, also as it is not intrusive the patient can wear it to work and in public without the social stigma that may come with wearing a large robotic device. The IPMC strips are soft and compliant therefore there is no danger of causing any harm to a patient.

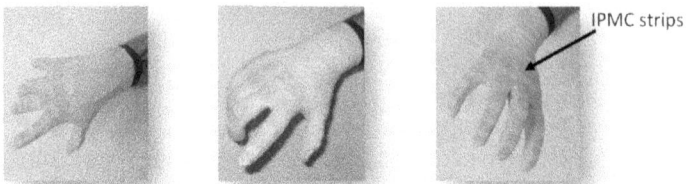

Fig. 6.2. Placement of IPMC strips at each finger joint for a proposed glove hand augmentation device.

It also envisaged that this design can eventually be developed into fully integrated lightweight human hand prosthesis for patients who have lost one or more of their fingers. With the full integration of a number of IPMC joints with sensing and electronics, as seen in Fig. 6.3, an elegant design can be achieved to mimic real finger joints.

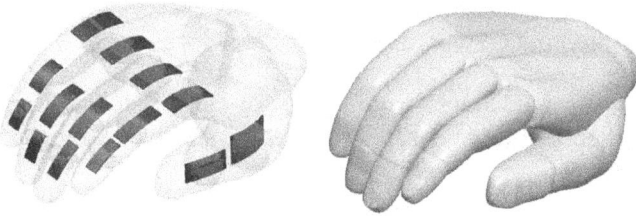

Fig. 6.3. Proposed hand prosthesis.

These exoskeletons and prostheses can be intelligently controlled using 'soft' cantilevered bending IPMC actuators and sensors and when coupled with electromyography (EMG) signals from the patient and embedded electronics a fully biomimetic hand device can be realized.

6.2. Gain Scheduled Nonlinear Control

6.2.1. Background

There have been many attempts by previous researchers to control the bending actuation of an IPMC transducer. LTI models have been used to develop LTI controllers. These have been implemented in a real system with limited success. The major problem, which was quickly realized, is that when you linearize the highly nonlinear and time-variant IPMC plant, the performance of the developed controller will dramatically decrease when the system operates further away from the equilibrium point, i.e. at high displacements. Therefore in order to accurately control the complex system over a large range of conditions more advanced controllers are needed.

The GS is a nonlinear control scheme commonly used in research and industrial applications such as flight control, vehicle control and automotive engine control. In traditional GS control, the nonlinear plant is linearized at a number of finite operating or equilibrium points. Linear controllers are then developed at each operating point, resulting in a set of linear controllers which exhibit good performance at each point. The control parameters for each of the linear controllers are then interpolated between their operating points, based on an appropriate design schedule. The resulting GS controller has the same architecture as the linear controllers but with parameters that are continuously varying with respect to the chosen schedule, resulting in a nonlinear controller. Two research papers [177, 178] present a history and in depth details about GS controllers.

The two major guidelines for the choice of the scheduling variables for the controller in order to ensure accurate and stable response are:

1. The scheduling variable should encapsulate the nonlinearities of the plant, and

2. The scheduling variables should vary slowly compared with the plant itself [179, 180].

A number of different techniques have been implemented in literature including; scheduling parameters on the reference trajectory and on the plant output [181]; scheduling the gain, poles and zeros of the controller transfer function [182] and; linear interpolation of the elements in a state space representation of a system [183].

The GS approach to designing a nonlinear controller has the major advantage of utilizing simple linear control techniques, which are extremely well understood and hence a large number of approaches exist. The main disadvantage is that no real performance or stability and robustness guaranties exist at present, except where the scheduled parameters are varying slowly [180, 183].

Most research into the control and modeling of IPMCs has been limited to small ranges of deflection in an attempt to reduce the nonlinearity inherent in the IPMC and improve the accuracy and repeatability of experiments. However if IPMCs are to be implemented and functional in real applications they must have excellent performance over a large operating region. It has been shown by Kothera [102] that it is advantageous to use different controller parameters if actuating to different target displacements. Therefore it was decided to implement a GS controller in order to accurately control the position of the IPMC over a large range of displacements, for both micro and macro movements.

The motivation for using this type of controller is that well understood linear control techniques could be utilized and also that the IPMC has been shown to be nonlinear based on the level of actuation [16, 149], lending itself well to a scheduled gain controller. The scheduling variable should then be based on the level of actuation (either voltage input or displacement output) to encapsulate the nonlinearities of the plant. Using the voltage input is not a good choice because when implemented, the controller output and hence IPMC input can change very rapidly to try and track a reference. The IPMC dynamics are relatively slow in comparison to these fast input changes therefore the second stability and robustness condition for the system, that the scheduled variable changes slowly, will not be guaranteed. It therefore seems appropriate to schedule the gains on the IPMC displacement, however there are issues surrounding this. Firstly the output may contain a significant amount of noise, especially when actuating the IPMC at micro targets, therefore the controller gains will vary rapidly and cause large jerky and even unstable oscillatory motions and secondly there will be implementation issues as the system would need to be tuned at a number of different outputs, but the actual output changes dynamically. A solution to this is described in the next section which details how the proposed GS controller will overcome these issues.

6.2.2. Proposed Gain Scheduled Controller

The proposed solution to this is to schedule the controller gains based on a function of the reference trajectory. The motive for this is that in position tracking applications, the objective is that the output signal follows the reference trajectory ($y \approx y_d$) and hence typically the output is just a low pass filtered version of the reference [167]. So actually the parameters will vary with the IPMC displacement, despite actually being scheduled on the reference signal. The second GS assumption of slowly varying control parameters will also hold true because as the controller is being designed for changes in reference set point and the parameters are scheduled on reference, which is a stair-step function, over the experiments the controller parameters will be piece-wise constant.

The designed GS controller is shown schematically in Fig. 6.4 with the associated variables. $G_C(z, \rho)$ represents the digital controller transfer function, G_{Joint} is the plant $f_{GS}(r)$ is a function that schedules the controller parameters, ρ, based on the reference input, $r(t)$, and outputs them to the controller. The signals $u(t, \rho)$ and $y(t, \rho)$ are the control effort and mechanical joint output, respectively, which are dependent on ρ. $v_u(t)$ represents the disturbances introduced at the controller output, for example electrical noise in the DAQ card and interface circuits, and $v_y(t)$ represents possible disturbances introduced at the output, for example mechanical vibration etc.

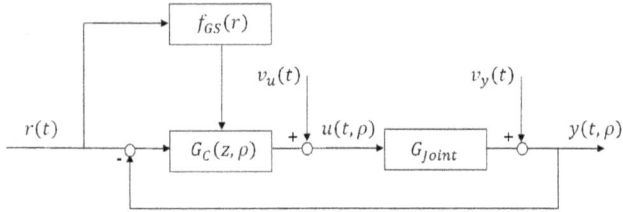

Fig. 6.4. Control diagram of proposed GS controller.

A PID control algorithm has been selected for the linear controllers. This was chosen for its simplicity to implement, ease of use with the IFT algorithms and because it has shown to provide good results for position control of IPMCs over a limited range [104] and as such should provide the basis for a stable and robust GS controller with good tracking performance. The steps to realize the GS controller are first, develop a set of linear controllers for different reference inputs across the desired operating range using IFT, next interpolate the gains K_p, K_i, and K_d, as a function of reference position, finally implement and test the controller in comparison with a standard linear controller.

6.3.   Experimental IPMC Results

The experiments were undertaken using the same test rig as shown in Fig. 5.4. A Nafion® based IPMC was used, with platinum electrodes. The IPMC was 35 mm long, 10 mm wide with a thickness of 200 μm. The clamped length was 5 mm. This relatively long length of IPMC was chosen because most research has been carried out with shorter lengths of IPMC as actuation response is more linear with shorter IPMCs [14, 131]. Therefore, to tackle the nonlinearity, a long length was used. Also shorter IPMCs cannot achieve large displacements so a long length will be needed to ensure that both micro and macro displacements can be investigated. With the specific IPMC used for this research, up to a 3 mm displacement will be input as the target reference.

There is an appreciable level of noise in the system, with a maximum ±30 μm mainly due to the laser sensor resolution and the quality of the reflected beam back to the laser sensor. A contribution to the noise may also be due to the control electronics and DAQ card. At displacements less than 200 μm there is a low signal-to-noise ratio and this will prevent accurate tuning at low displacements because of the lack of information obtained for the IFT algorithm. IFT still tunes the controller, but the results are inconsistent and are highly dependent on the noise contribution. Despite the inability to reliably tune at these displacements, the controller can still accurately track to these targets but the output signal includes a noise error of ±30 μm. This is shown in Fig. 6.5 where the controller can accurately track to the 100 μm target, but because of the level of noise in the system (approximately 1/3 of the target value), tuning at this level is inconsistent.

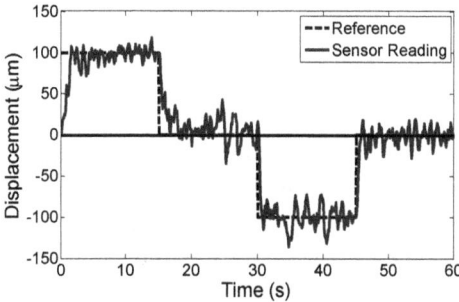

Fig. 6.5. Typical tracking of a 100 μm target.

The same settings for the initial controller gains and IFT step size that were used in the previous chapter have been used for the GS controller, i.e. initial controller gains are, $K_p = 1000; K_i = 500;$ and $K_d = 0$, step size $\gamma_{K_p} = 1$; $\gamma_{K_i} = 1$; $\gamma_{K_d} = 0.5$ and sampling time of 0.1 s.

6.3.1. Tuning for Different Target References

Due to the desired applications in biomedical robotics, the control system was designed to be accurate for changes in set point, in terms of both the transient and steady state response. The same tuning procedure for IFT as described in Section 5.4 was undertaken for 100 μm, 200 μm, 300 μm, 500 μm, 1 mm, 1.5 mm, 2 mm, 2.5 mm and 3 mm reference displacements. This reference will ensure that the IPMC has been tuned in both directions, as it has been shown that due to imperfect fabrication techniques the IPMC can have different performance in different directions, and will also tune for 4 transient periods as well as steady state behavior.

It was found that at 100 μm and 200 μm despite being able to track accurately to the target displacements, the noise level from the sensor and electronics was too large to accurately tune for these displacements. The tuned parameters were very inconsistent because of the low signal-to-noise ratio. This is confirmed as the IFT algorithm tunes the proportional gain to zero or very low for these displacements to ensure that the noise is not amplified.

The controller gains and design criteria are plotted in Fig. 6.6 for a small, 300 μm; medium, 1.5 mm and large, 3 mm displacement for each iteration to show how they are being dynamically updated through this tuning algorithm. It can be seen that the design criteria converges to an optimal solution in all cases.

The full set of results for the final tuned values for each displacement is given in Table 6.1, along with the percentage improvement of the controller design criterion, *J*, in equation (5.27). The shaded results in the table represent the experiments in Fig. 6.6 (a), (b) and (c). It is very clear to see the major success of the IFT algorithm in tuning the system from an arbitrary controller without the need of any plant model.

Table 6.1. Summary of the results for IFT at different target displacements.

Target Displacement (mm)	Final K_p	Final K_i	Final K_d	Initial J $(\times 10^{-9})$	Final J $(\times 10^{-9})$	% improvement
0.3	3092.56	2813.93	273.26	3.2411	1.9618	39.47
0.5	2636.35	2791.43	483.04	11.383	3.9523	65.28
1	2019.93	1904.23	436.34	29.289	12.650	56.80
1.5	1480.53	1903.93	476.95	58.403	26.322	54.93
2	1488.18	1585.71	539.94	91.605	52.129	43.09
2.5	1577.95	1128.05	708.01	296.87	113.38	61.81
3	980.449	1040.13	354.75	288.56	127.93	55.66

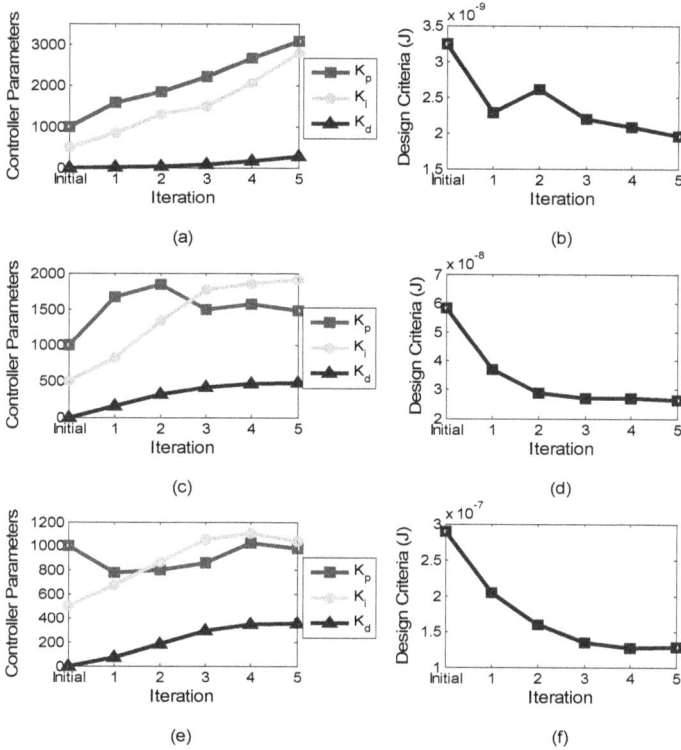

Fig. 6.6. Controller parameters and design criteria after each iteration for (a) 300 μm, (b) 1.5 mm and (c) 3 mm reference input.

The results for the controller parameters are plotted in Fig. 6.7 to display how the gains change after each iteration and at each target displacement. If a smaller or larger step size, γ, for the IFT size was chosen these plots would be either smoother or more bumpy, respectively. The effect of target displacement on the tuning of the control parameters can clearly be seen as the initial controller for all reference displacements are the same, yet the controller tunes to very different states for the different reference inputs.

6.3.2. Development of the Gain Schedule

The schedule for the controller gains has been chosen based on a function of the reference trajectory. Its architecture is shown previously in Fig. 6.4 and the task now is to find the function, $f_{GS}(r)$ for the controller.

(a)   (b)

(c)

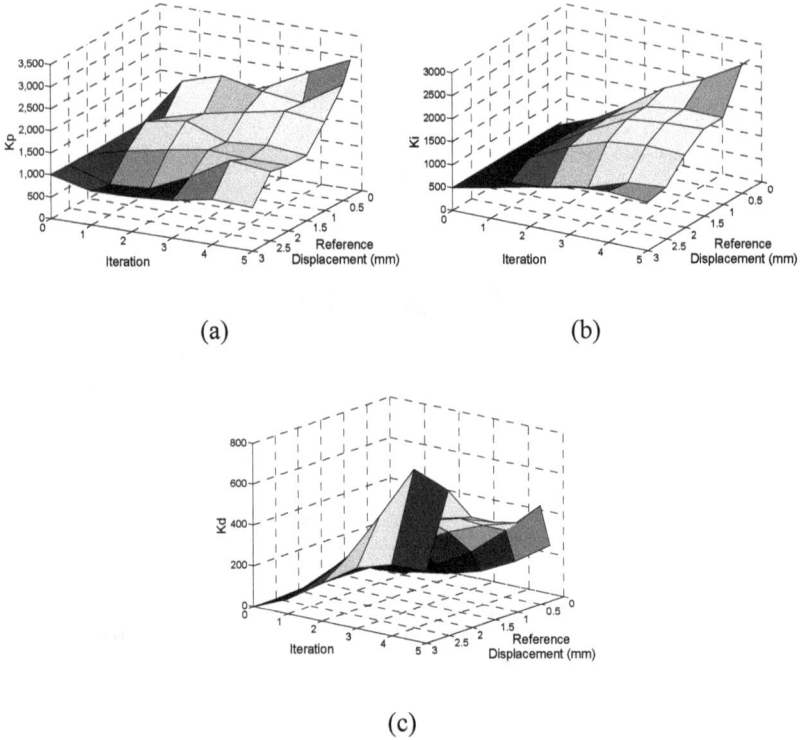

Fig. 6.7. (a) K_p, (b) K_i and (c) K_d, plotted for each tuning iteration, as a function of target displacement.

The IPMC PID controller has been tuned using IFT for a range of different target values, resulting in a set of tuned linear controllers for different reference trajectories. Now in order to turn these finite linear controllers into a continuous controller, the control parameters are interpolated in order to schedule the gains continuously over the operating range. The tuned controller gains K_p, K_i and K_d are plotted in Fig. 6.8 as a function of target displacement. There is a clear trend in the relationship with the control parameters and target displacements. After some analysis it was found that a logarithmic fit would give the best correlation between these parameters. These trends for the control parameters are plotted on the graph below and the relationships are presented in equations (6.1), (6.2) and (6.3).

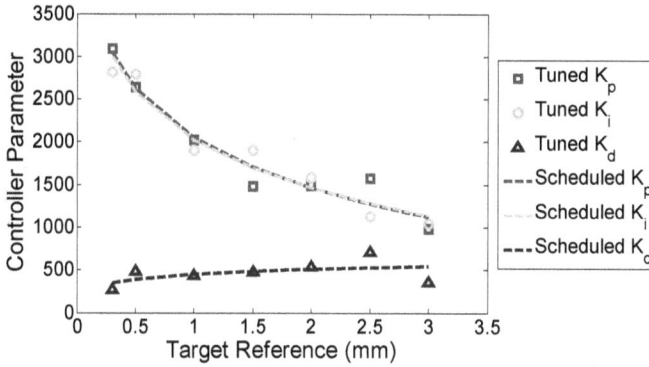

Fig. 6.8. Final controller parameters after IFT at varying displacements.

$$K_p = -839.7 \ln(r) - 3757.6 \qquad (6.1)$$

$$K_i = -808.0 \ln(r) - 3560.2 \qquad (6.2)$$

$$K_d = 84.73 \ln(r) + 1038.1 \qquad (6.3)$$

The controller was also tuned for references of 100 μm and 200 μm but because of the low signal-to-noise ratio the parameters were very inconsistent. The tuned values for K_p and K_i at 100 μm and 200 μm were in the region of those obtained for 300 μm and the values for K_d were slightly lower than that for 300 μm at these small targets. For this reason it was decided to restrict the controller parameters at their 300 μm values i.e. if the target displacement was less than 300 μm then use the values scheduled for 300 μm. This is also necessary to prevent K_p and K_i from increasing to an extremely high level because their schedule approaches the zero reference asymptotically.

An interesting observation is that if the IPMC was linear, then the controller gains would be constant across all target displacements. This in itself validates that the IFT algorithm has successfully tackled the nonlinear characteristics of the IPMC. It can also be observed that the IFT algorithm realizes that the signal-to-noise ratio is low at small displacements and consequently reduces K_d accordingly as not to amplify the noise at these levels.

Now the schedule to change the gains has been developed, $f_{GS}(r)$, the nonlinear GS controller is complete and ready for implementation.

6.3.3. Comparison of GS Controller with Linear Controller

The developed nonlinear GS controller was tested for a number of different reference trajectories and its performance compared to a conventional PID controller. It was decided to use the tuned controller parameters for 1.5 mm displacement for the PID controller, as this is in the middle of the range of IPMC operation. The experimental design criteria for IFT, equation (5.27) was used as a quantitative measure of the performance of the controllers and qualitative measures such as overshoot and settling time are also analyzed. It was clear that after tuning the GS controller for different step inputs, as per Table 6.1, that the GS controller would outperform the PID controller for all step target displacements except for at 1.5 mm, where the two controllers will be the same. Consequently no comparisons for the step inputs are presented.

In order to test the performance of the GS controller versus the conventional PID controller for various changes in set point, which is what the controller is designed for, a random stair-step sequence of varying amplitude was input. Both the micro and macro targets were used. The results of this test are shown in Fig. 6.9. By examining the plot it can be seen that the GS controller had a much smaller overshoot at all of the set point changes, except for at micro targets (30 s and 90 s). This is due to the fact that the proportional gain is high when the target displacement is low as seen in Fig. 6.8. This suggests that the cut-off region in the schedule, which was set at 300 µm, may be too low if small overshoots are required for a specific application. Using a higher cut-off (say 500 µm) will restrict the K_p and K_i values at micro displacements and the GS controller may not over shoot as much, but this will no doubt result in longer rise times. The settling time after a set point change is better for the GS controller in all cases, even at micro level targets, where there is more overshoot. Comparing the overall error for the controllers using the design criteria, $J(PID) = 1.19e^{-7}$ and $J(GS) = 1.02e^{-7}$ the GS controller is approximately 17 % better.

In order to demonstrate the versatility of the GS controller to other input signals, sinusoid reference trajectories were tested versus the conventional PID controller. This will result in dynamically varying control parameters as the reference signal is continuously changing. A number of experiments were undertaken to assess the performance under different conditions inside the desired operating range of the

IPMC. This will also test the robustness of the GS controller with respect to the GS guideline that the scheduled parameters must vary slowly.

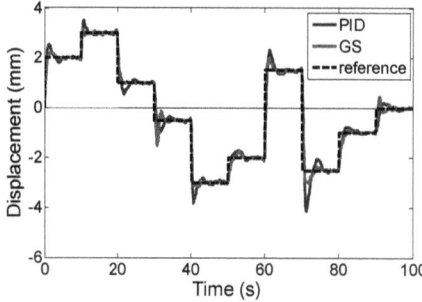

Fig. 6.9. Random stair-step reference input for comparison of controller performance.

Fig. 6.10 (a) shows the 33×10^{-3} Hz signal with a micro amplitude of 500 μm. It can be seen that both controllers follow the reference very well, despite a relatively high level of noise. By inspection the performance of both controllers are comparable, but using the design criteria, $J(PID) = 1.32e^{-9}$ and $J(GS) = 1.15e^{-9}$ it can be seen that the GS controller does perform better.

Fig. 6.10 (b) shows the 33×10^{-3} Hz signal with a large amplitude of 3 mm. Both controllers track the reference extremely well and again by inspection the performance of both controllers are comparable. Using the design criteria, $J(PID) = 6.55e^{-9}$ and $J(GS) = 5.35e^{-9}$ it can be seen that again the GS controller does perform better.

It has been shown that the designed controllers can accurately track a dynamic reference with a time period of 30 s, so a faster signal was tested to push the guideline of slowly varying parameters. A time period of 10 s should be tested as this is nearing the limit of the speed of the IPMC. Fig. 6.11 (a) shows the 0.1 Hz performance at a 500 μm amplitude for the two controllers. It can be seen that the standard PID controller has a considerable level of overshoot and then consequently lags the reference signal. It is clear that the GS controller is performing better and this is confirmed by the design criteria, $J(PID) = 16.3e^{-9}$ and $J(GS) = 2.35e^{-9}$.

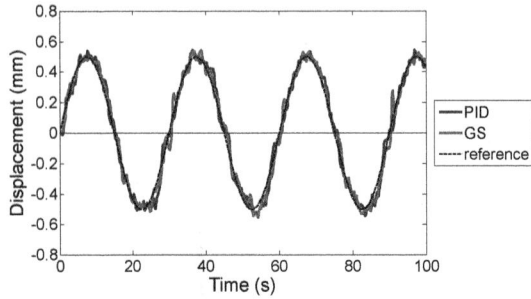

(a)

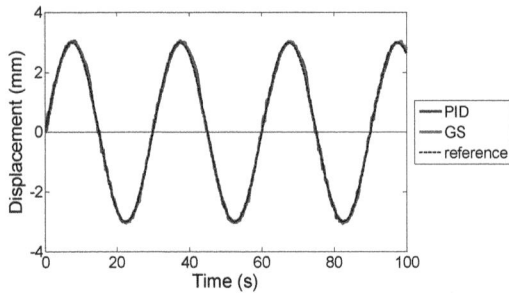

(b)

Fig. 6.10. 33×10^{-3} Hz sinusoid inputs for (a) 500 μm
and (b) 3 mm amplitude.

Fig. 6.11 (b) shows the performance with reference amplitude of 3 mm at 0.1 Hz. Similar to the 500 μm reference it is clear to see that the standard PID controller exhibits overshoot. This high level of overshoot again results in an output lag. By inspection the GS controller performs a great deal better and this can be confirmed by the design criteria, $J(PID) = 13.6e^{-8}$ and $J(GS) = 6.96e^{-8}$.

6.4.   Mechanism Experiments

The GS controller has been proved to work extremely well on the IPMC, achieving accurate displacements over a large range and at low frequencies, where the IPMC is most nonlinear. This has validated that the GS controller is useful for the rotary joint mechanism which was designed in Section 6.1. The mechanism and integrated IPMC actuator are   assembled   and   open-loop   experiments   are   undertaken   to characterize the system. The results are shown in Fig. 6.12, with the IPMCs by themselves as well as when integrated with the mechanism.

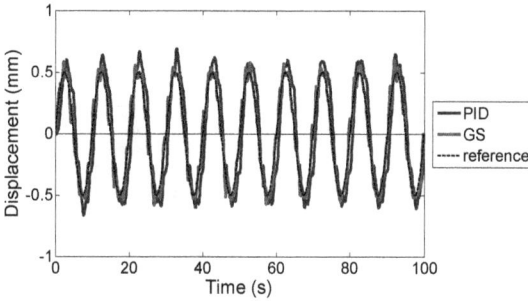

(a)

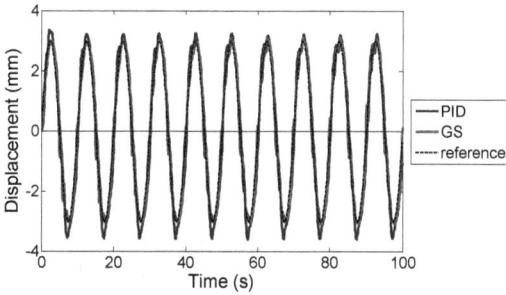

(b)

Fig. 6.11. 0.1 Hz sinusoid inputs for (a) 500 μm and (b) 3 mm amplitude.

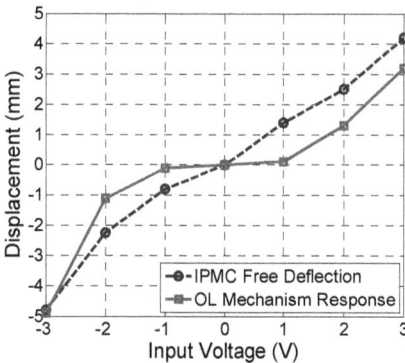

Fig. 6.12. Open-loop experimental IPMC and mechanism response.

119

As the saturated output for the IPMC is chosen at ±3 V, the desired average operating voltage for the IPMC under closed loop control is around 2 V. The model developed for the IPMC actuated rotary mechanism is simulated for a 2 V input and the results are shown in Fig. 6.13. The model and simulations show excellent correspondence.

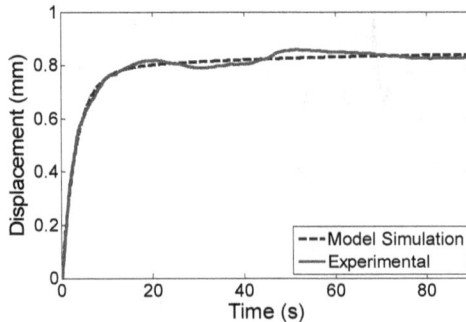

Fig. 6.13. Simulated and experimental IPMC and mechanism response.

From the results in Fig. 6.13 it can be seen that only displacements in the micro range can be achieved with the IPMC sample, which is used in the example to drive the mechanism. As such the desired reference for the IFT on the mechanism will be restricted to 300 μm as this is the smallest reference which was accurately tuned for in the GS controller. As the IPMC is not strong enough to drive the mechanism arm into the high displacements the GS controller cannot be tested on the mechanism and hence only the IFT tuning algorithm can be validated.

The mechanism is tuned from three initial states, model-tuned, over damped and under damped, to show the versatility of the IFT to tune the device from various states. It is clear in all cases that the IFT improves the system response, even from a controller that was tuned using the model of the IPMC system as shown in Fig. 6.14.

6.5.   Discussion

IFT is based on the optimization of a specific control design criteria. It has been shown that for a time invariant system if the IFT step size and starting controller are chosen appropriately the system will converge to this optimum state. As IPMCs are time-varying the plant dynamics will change between and even within an iterative experiment. As such it is

not possible to confirm that the controller parameters will converge to
an optimal state, as the entire system is constantly changing. To
overcome this, tuning must be either done online, throughout the
normal operation or must be undertaken at timely intervals, i.e. when
the IPMC dynamics vary so much the current controller performance
no longer meets the design specifications, the system should be tuned
again.

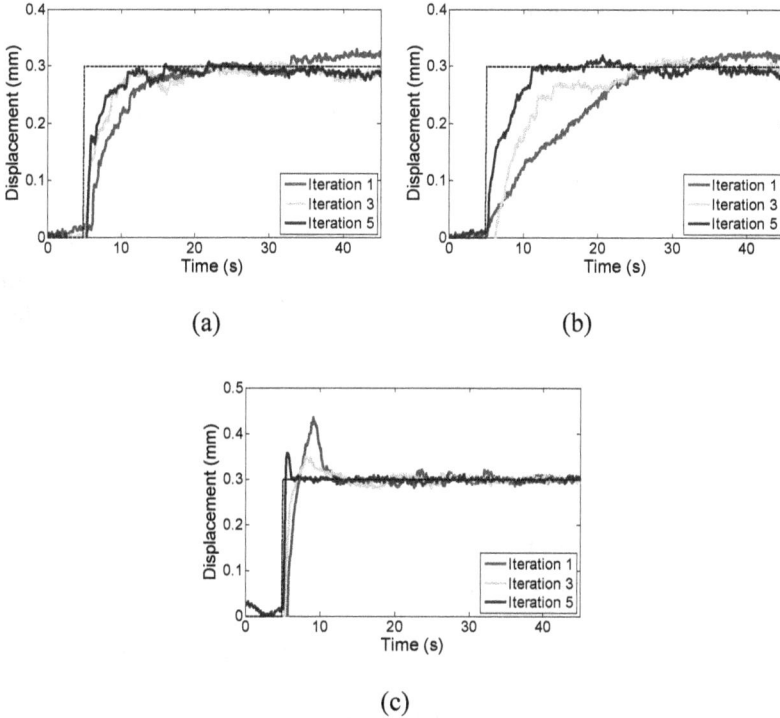

(a)   (b)

(c)

Fig. 6.14. Experimental IPMC and mechanism response tuning from an initial
(a) model tuned, (b) over damped and (c) under damped controller.

The IPMC system was tuned for 5 iterations only. This was done to
restrict the number and time of experiments to try and keep the system
as time invariant as possible, as the IPMC performance does fluctuate
with time. If the system had been tuned over more iterations then a
controller which is closer to the optimal may have been achieved,
although Fig. 6.6 (a), (b) and (c) demonstrate that the design criterion
converges very quickly after only 5 iterations. It should also be noted
that the gradient search algorithm used for this implementation of IFT

will converge to a local minimum. If the initial controller was at another state far away from the one used, then the controller may have converged to a different minimum resulting in an alternate optimally tuned controller.

The final GS controller exhibits overshoot to set point changes as a trade-off with settling time. This characteristic of the system can easily be altered by placing a weighting filter in the design criteria when tuning the controller to ensure more emphasis on either the transient or steady state performance.

Here a specific size of IPMC was used and the rotary mechanism was designed to fit this. As an IPMC actuator was chosen in advance the task was to reduce the weight and friction in the rotary joint so the IPMC could drive it. The design was successful as the IPMC can move the arm, but only micro displacements are achievable when driving the mechanism. If the arm was prototyped from a more accurate material the friction may have been reduced but this would have increased the weight of the mechanism. As the state of art in IPMC manufacture increases the available force output of the actuators will increase and thus if a new IPMC were available which could drive the mechanism to high displacements, then the GS controller could easily be implemented on the system, just as it was on the IPMC by itself.

6.6. Rotary Finger Joint and Iterative Feedback Tuning Gain Scheduled Summary

A rotary mechanism with an integrated IPMC actuator has been designed to mimic a biological system. The biomedical robotic joint has been developed and implemented. This is one of the first examples of using an IPMC to drive a real mechanism, with friction, gravity effects etc. This is also one of the few devices that uses closed loop control on an actual real life device. An extension of the design to a full human hand exoskeleton and prosthesis has also been proposed.

A GS controller has been developed using PID architecture with the controller parameters, K_p, K_i and K_d scheduled on the reference trajectory of the mechanism. This was done to improve the IPMCs performance over a large range of displacements, both in the micro and macro range, because the nonlinearity of the IPMC is based on actuation level. The performance of the GS controller was tested against a standard PID controller and has shown improved performance at large and small deflections and at different frequencies.

The IPMC used to drive the mechanism can only achieve displacements in the micro range and as such a GS controller could not be tested on the mechanism itself. IFT has demonstrated good tuning results in tuning the mechanism in this micro range and the GS controller implemented on the IPMC has demonstrated excellent results, validating both the IFT and GS methods. If a new thicker and hence more powerful IPMC actuator was to be used and higher mechanism displacements could be achieved, it is reasonable to assume that the GS controller would work equally as well as it has on the IPMC by itself.

Chapter 7

Microfluidic Pump
with Online IFT Control

One promising application of Ipmcs is in the delivery of medical drugs through the development of appropriate microfluidic pumps. Traditional intake of drugs for patients with tablets cause a large dose soon after the intake followed by decay in the medicine over time before the next dose is taken. This situation is not ideal as the dosage should be more stable overtime. Intravenous therapy delivers drugs directly to the veins and can give a more stable dosage of medicine. The problem is that the standard equipment required is not portable and consequently this is primarily carried out in hospitals. A solution to this is to develop a portable micropump for automatically dispensing drugs at a more constant rate. The problem with this is in the miniaturization of the drug dispensing system, pump, reservoir, embedded electronics etc. IPMCs lend themselves well to miniaturization and can be used as an integrated actuator in an embedded pump-reservoir system for the microfluidic device which can operate for long periods of time due to the low power consumption.

One key specification for such a micropump is to operate accurately for long periods of time to ensure stable drug delivery. This is a major issue for the proposed system as IPMCs are extremely time-varying. Adaptive control systems must be implemented to accurately control the displacement and hence the dispensing rate. Previous methods which have attempted to control IPMCs over a period of time are very limited and papers with convincing experimental results are scarce. In [39], a MRAC scheme was implemented with a genetic algorithm approach to optimize the controller. Limited simulation results are presented and there are no experimental results to verify that the algorithm does actually improve tracking performance over a long period of time. Another MRAC is implemented in [40], where a LTI model is used as the reference. Experimental results are presented but only over a relatively short period of time, up to 4 minutes in one experiment. Many standard adaptive controllers may not cope when the

IPMC dynamics vary too far from their initial state and hence any developed reference model.

IFT is a model-free tuning algorithm which automatically adapts the system parameters towards an optimal system performance and hence is very suitable for a highly nonlinear and time varying system such as IPMCs. The standard IFT algorithm as described in Chapter 5 requires a 'special' gradient experiment which restricts many systems from being tuned online during normal operation. The research described in this chapter develops a new online IFT approach which allows the system to be continuously tuned during normal operation, throughout the life of the device, where the standard IFT requires the device to be taken out of use for it to be tuned. This chapter explains how the new IFT approach has successfully tackled the time-varying characteristics of the IPMC which will permit new applications, specifically in microfluidics.

The development of the IFT algorithm and results has been published in [185] and experimental results on the micropump are to be presented at [186].

7.1.   Micro-pump Application

Micro-pumps are the key components in microfluidic systems. They are widely used in many applications such as lab-on-chip, micro-total analysis systems and micro-dosage systems. The first micropump was fabricated in the 1980s, a recent review of current micropump technology can be found in [187].

The most important component in a micropump is the actuating mechanism because it is directly related to factors such as flow rate, driving source and cost. Currently the most common driving mechanisms are piezoelectric, electromagnetic, thermo-pneumatic and electrostatic. They all have their own advantages and disadvantages. Most are relatively expensive with complex fabrication processes. In this research a low-cost micropump design with simple fabrication process using simple CNC machining is developed. The micropump is driven by an IPMC actuator. The main advantages of using IPMCs as the actuating mechanism is they have large displacements, which will correspond to large flow rates, at low voltages, which makes them good for long-life in portable and embedded applications. IPMCs have been shown to operate better in reciprocating motion rather than DC actuation. IPMCs can also easily be scaled down for micro applications making them easy to implant as they are also biocompatible with humans. The main challenge is that IPMCs are not very stable actuators and their performance varies significantly over time. Here, we demonstrate a PID controller which is continuously tuned using a new IFT algorithm in order to overcome this challenge and ensure a controlled pumping rate over an extended period.

7.1.1. Mechanical Design

Fig. 7.1 shows the design for the valve-less micropump. It consists of four Perspex layers (each 2mm thick except the bottom layer) and a latex diaphragm. The top layer is the cover. It also has a micro-channel inlet, which can be connected to an external reservoir. The second layer is the most critical part which contains inlet/outlet ports and the pump's micro-channel outlet. It requires precision machining with very low tolerances to ensure the diffuser and nozzle dimensions are correct for proper operation of the micropump. The third Perspex layer holds the pump chamber. A latex film is directly attached, glued, to the third layer to cover the pump chamber at the bottom. A small piece of strong double sided tape (1 mm × 1 mm) of about 1 mm thick is placed underneath the latex film at the center where the IPMC tip is attached to the latex diaphragm. The IPMC contact area via the double sided-tape to the diaphragm is small; hence exerting a high pressure for a given force (approximately 1 gf) of the IPMC. To overcome the dehydration problem, the IPMCs should always be kept in a liquid. A bottom layer with a cavity to hold water is added. This layer is thicker than the other layers (4 mm instead of 2 mm) so that the cavity is sufficient to hold water while providing enough space for the IPMC strip to deflect. In this prototype the bottom layer also houses the two metal electrodes which clamp each side of the IPMC. The clamped section of IPMC is also submerged in water, thus continual operation is possible. Since this is a prototype only, the pump was not made to be very small, but the system is scalable for miniaturization after the pump concept has been proved. The overall size is 30 × 40 × 10 mm (length × width × height). The pump chamber is a square of 11 × 11 mm with a depth of 2 mm. The actuator used in this prototype is a Nafion®-based IPMC strip with platinum electrodes. The IPMC dimensions are 17.5 mm long, 10 mm wide with an average thickness of 200 μm. A pair of copper electrodes was used to clamp the two sides of the IPMC strip, providing electrical stimulus to the IPMC. The clamping length is 5 mm.

The channels, nozzles and diffusers were CNC micro-machined using a milling tool. The channels are 500 μm wide and 300μm deep, while the conical nozzles/diffusers have the dimensions of D_0 = 100 μm, D_1 = 800 μm and L= 2 mm. With these dimensions a ratio of flow resistance coefficients of nozzle to diffuser, η of about 3 can be achieved [188].

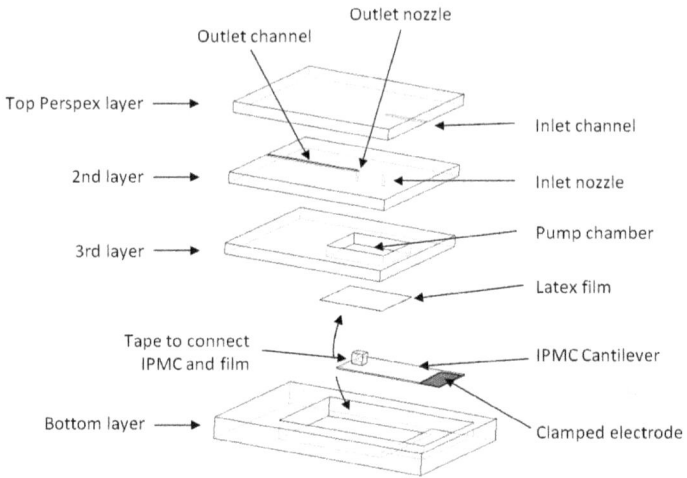

Fig. 7.1. Exploded view of micropump, designed for dispensing drugs. [185].

7.1.2. Modeling and Simulation

The IPMC model is used with a simulation of the pump performed using ANSYS®, finite element analysis software and its CFD program called ANSYS® CFX. Fig. 7.2 shows the simulation results for a complete pumping cycle with the conditions of P = 1200 Pa and f = 0.2 Hz. The period for one pumping cycle is 5 s; thus 50 time steps. Two simulation frames which correspond to the pump and supply modes are shown in Fig. 7.2 (a) and (b) respectively. The red colour corresponds to high pressure. In Fig. 7.2 (a), fluid starts to flow out from the chamber towards the inlet and outlet channels as a result of the IPMC pushing the latex diaphragm up. It can be seen there is a lot of fluid travelling towards the outlet channel and there is only a little fluid flowing towards the inlet channel. This is correct operation because the diaphragm is being pushed upwards which corresponds to the "pump mode". Fig. 7.2 (b) illustrates the simulation result when the diaphragm is moving downwards. This time more fluid flows from the inlet channel into the pump chamber than fluid flowing out of the outlet channel, corresponding to the "supply mode". During simulation, the displacement of the diaphragm follows a sinusoidal wave with a maximum displacement of 50 µm.

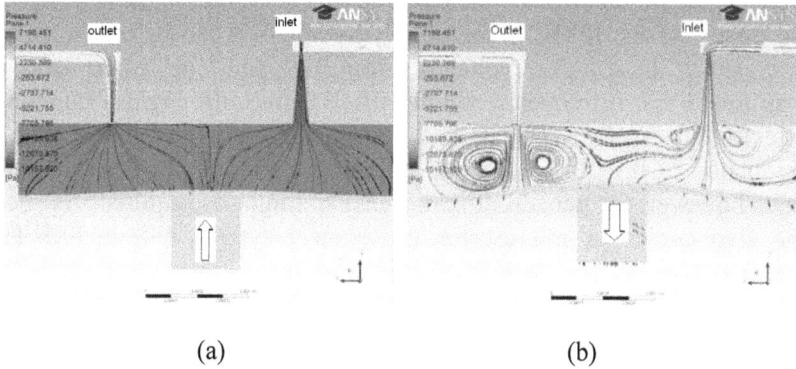

(a)   (b)

Fig. 7.2. ANSYS simulation for (a) pump, and (b) supply mode.

7.1.3. Prototype

The constructed prototype of the micropump is shown in Fig. 7.3. The clamps are in place to hold the pump layers together. The pump has not been completely glued as this is a prototype and it will need to be disassembled and assembled a number of times.

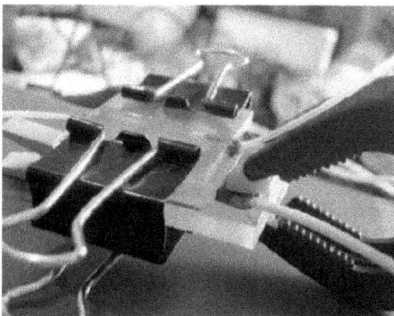

Fig. 7.3. Micropump prototype.

129

7.1.4. Open-loop Characterization

To realize a continuous pumping rate a sinusoid IPMC displacement is required. The performance of the micropump was initially tested with open-loop inputs to characterize the real system. Sine wave voltages of varying amplitudes and frequencies were applied to the IPMC and the average flow-rate is plotted in Fig. 7.4. The flow rate is higher at lower frequencies for peak voltages >2 V. This suggests that at lower frequencies a higher IPMC displacement is achieved, causing a higher flow rate. As the IPMC is continually hydrated, it can be continuously operated at these higher voltages. From the open-loop experiments it was found that a frequency around 0.1 Hz gives the highest pumping rate for this system. This is in line with [126], which shows that a frequency of 0.1 Hz gives good pumping operation. Lower pumping rates however, for general drug dispensing, might be required. As such the frequencies of interest for the control system will be 0.1 Hz and 0.05 Hz.

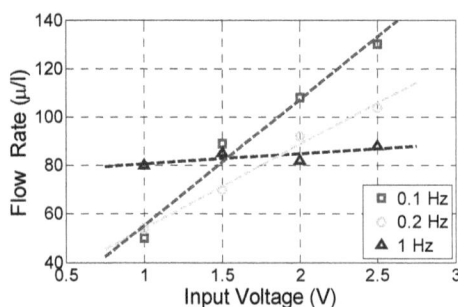

Fig. 7.4. Average flow rate for varying amplitude voltage sinusoids at different frequencies.

7.2.   Control and Tuning of Micropump Actuating Mechanism

7.2.1. Control System

In order to effectively control the IPMC displacement a standard 1DOF PID control architecture is implemented, as shown in Fig. 7.5. $G_C(z, \rho)$ is the controller, ρ is the controller parameter vector which contains the parameters to be tuned, $\rho = \begin{bmatrix} K_p K_i K_d \end{bmatrix}^T$ and G_{pump} is the plant (model for simulations and actual IPMC and pump for experiments). *r(t)* is the reference input, $\tilde{y}(t)$ is the position error, $u(t,\rho)$ is the voltage input to the IPMC and $y(t,\rho)$ is the output position. The disturbances into the system at the pump input and output are represented by the signals, $v_u(t)$ and $v_y(t)$, respectively. Typical disturbances may include electrical noise, vibrations when the pump is moving, sensor noise etc.

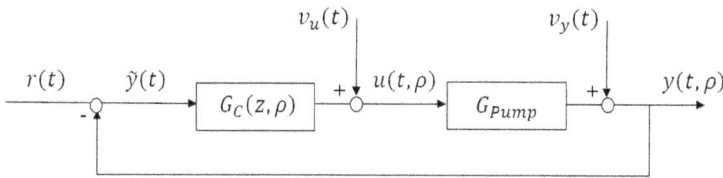

Fig. 7.5. Closed loop control system for IPMC micropump application.

This approach is easily implemented in real-time applications as no model is required to develop this controller. It has been demonstrated in [19, 37, 46, 104] that simple linear PID control can in fact accurately control the IPMC system, operating within a certain limited range. It is also shown in [150] and validated in [16] that IPMCs are more nonlinear at high displacements and low frequency to DC. This can therefore be used as a justification for implementing a linear controller for a system with micro displacement and with periodic signals (i.e. >DC) such as microfluidics, the target application for this IPMC research.

The PID control system is also the most commonly used control system for industrial applications and therefore the application of IFT to this system is valuable research which can be transferred for the online automatic tuning of any industrial system; even for applications which are nonlinear and time-varying, as most real life systems are.

7.2.2. Controller Tuning

Most IPMC controllers that have been realized are tuned in simulation using an approximate plant model. The performance is then assessed, also in simulation, before implementing the controller on the real system. This controller must be further fine-tuned on the real system to account for variability between the model and the real plant. This method for developing controllers is also sample specific as a tuned controller cannot be transferred to a different IPMC actuator due to large variability in fabrication and hence performance in all IPMCs [103]. Also the fact is that IPMC dynamics do vary far from their initial state and the performance of a static model-based controller can become unacceptable. If the system dynamics of an IPMC drift far from the developed model then the effort spent on modeling the system becomes redundant. Consequently it is highly desirable to develop an automatic tuning method. The IFT method developed here can be used as for automatic online tuning of any system without the need for any model or knowledge of the system. This is highly desirable as even though some model-based control methods have shown reasonable performance they have only been proven to operate well only over a short period of time [39, 40].

Since a model has been previously developed in Chapter 3, it is advantageous to use this model to find a stable controller as a starting point for the IFT routine. This model tuned controller will also be used to measure the performance of the IFT algorithm, demonstrating its adaptive ability in contrast with a non-adaptive model-based control system. In this way the IFT algorithm will take over the model-based approach and automatically tune the system towards an optimal state even as the system dynamics change. The IFT algorithm itself does not use any model; the model is simply used to find an initial stable controller.

The material parameters in the model are updated for the new IPMC sample and as the model is geometrically scalable the new geometrical parameters were input in order to accurately model the new IPMC. The model does accurately represent the IPMC response at the time of experimentation, but over a period of continuous operation the highly time-varying nature of the IPMC will cause the response to change unpredictably. This cannot be modeled as the variance is due to ion redistribution which is a stochastic process. As such the simulations will give an indication of the system performance but it can be expected that the experimental results may vary significantly from the simulation results. This will show that any model-based controller tuning may result in a good system performance at the time of modeling, but cannot guarantee this performance at some later stage in the system operation. This is the motivation for the online automatic tuning algorithm.

7.2.2.1. Model Based Tuning

The first stage in developing a model based control system, after a model for the system has been found, is to tune the system based on simulation in order to obtain some acceptable control performance. The closed loop Ziegler-Nichols tuning method [189] is a well-known and commonly used rule based tuning method for PID controllers in industrial applications, where PID gains are calculated based on K_u and T_u, the critical gain and time period respectively. The conventional Ziegler-Nichols parameters for a PID controller are designed to give a quarter decay ratio response to load disturbances [189] and therefore oscillatory transient responses to set-point changes typically result from this rule. As will be discussed later in the next section of this paper, unwanted excitations may result from the IFT if the system starts from an initially under damped state and as such a modified Ziegler-Nichols rule-based tuning method which is presented in [190] is used. This method is a modification on the standard Ziegler-Nichols method which attempts to eliminate the overshoot to changes in the set point. The proportional gain is reduced by a factor of three and the derivative time increased by the nearly the same factor from the standard Ziegler-Nichols parameters, see equation (7.1).

$$K_p = 0.2K_u; K_i = \frac{2K_p}{T_u}; K_d = \frac{K_p T_u}{3} \qquad (7.1)$$

Simulation was undertaken to calculate the critical gain and time period, K_u and T_u, by setting the integral and derivative gains to zero and then slowly increasing the proportional gain to reach the critical point, where sustained periodic oscillations occur for a set point reference. It was found that $K_u = 8200$ and $T_u = 2$s. This results in a tuned system with gains of $K_p = 1,640$; $K_i = 1,640$ and $K_d = 1,093.33$.

Applying these values to the control system and simulating the IPMC model gives the following results for 0.1 Hz, 300 μm amplitude as seen in Fig. 7.6. The IPMC model is time invariant and therefore once the initial transient dynamics have passed the system will perform the same indefinitely if there are no major external disturbances on the actuator.

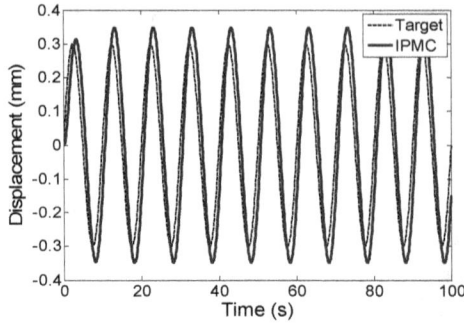

Fig. 7.6. Simulated IPMC output for modified Ziegler-Nichols tuned system, frequency=0.1 Hz, amplitude = 300 μm.

7.2.2.2. IFT Extension to Online Tuning

IFT has traditionally been used to tune systems offline, or before they commence standard operation due to the fact that a 'special' gradient experiment, whose trajectory may deviate far from the trajectory of the normal experiment, is needed to calculate the updated parameters.

The basic IFT algorithm for tuning a 1 DOF PID controller consists of two experiments, a normal and gradient experiment, as explained previously. Using this procedure means that for half the time that the system is tuning, the reference input is the error signal, and so the output will vary far from the normal or desired trajectory. In a practical sense, for online tuning it is not acceptable that the gradient experiment differs far from the normal experiment.

It was briefly suggested in [167] that it is possible to improve performance by using a different second experiment to calculate the gradient. The IFT algorithm in this research uses this idea to overcome the production waste caused by the second gradient experiment by 'switching' the reference for the gradient experiment. Steps 2 and 3 of the standard IFT algorithm outlined in Section 5.3.2 are now changed in this online implementation. In step 2 the reference for the gradient experiment, r^2, is changed to $y^1(\rho)$, the output from the first experiment, instead of $(r - y^1(\rho))$. So now $y^2(\rho)$ is now expressed as

$$y^2(\rho) = T(\rho)(y^1) + S(\rho)(Pv_u^2 + v_y^2) \qquad (7.2)$$

Step 3 of the standard IFT algorithm is modified so the estimate of the gradient $\partial y(\rho) / \partial \rho$ is taken as in equation (7.3) instead of equation (5.14):

$$\widehat{\frac{\partial y}{\partial \rho}}(\rho) = \frac{1}{C}(\rho)\frac{\partial C}{\partial \rho}(\rho)\left[y^1(\rho) - y^2(\rho)\right] \qquad (7.3)$$

where y^1 and y^2 are the closed loop outputs from the first and second experiments respectively. Now combining equation (7.2) with (7.3) and comparing with equation (5.10),

$$\widehat{\frac{\partial y}{\partial \rho}}(\rho) = \frac{1}{C}(\rho)\frac{\partial C}{\partial \rho}(\rho)\left[y^1(\rho) - T(\rho)y^1 + S(\rho)\left(Pv_u^2 + v_y^2\right)\right]$$

$$\widehat{\frac{\partial y}{\partial \rho}}(\rho) = \frac{1}{C}(\rho)\frac{\partial C}{\partial \rho}(\rho)\left[T(\rho)r - T(\rho)y^1 + S(\rho)\left(Pv_u^2 + v_y^2\right)\right]$$

$$\widehat{\frac{\partial y}{\partial \rho}}(\rho) = \frac{\partial C}{\partial \rho}(\rho)\left[PS(\rho)\left(r - y^1(\rho)\right) + \frac{1}{C}(\rho)S(\rho)\left(Pv_u^2 + v_y^2\right)\right]$$

$$\widehat{\frac{\partial y}{\partial \rho}}(\rho) = \frac{\partial y}{\partial \rho}(\rho) + \frac{1}{C}(\rho)\frac{\partial C}{\partial \rho}(\rho)S(\rho)\left(Pv_u^2 + v_y^2\right)$$

$$= \frac{\partial y}{\partial \rho}(\rho) + w(\rho)$$

This is the same result as with the standard IFT algorithm and so again when comparing with the actual gradient, it can be seen that there is a perturbation, $w(\rho)$, introduced due to the disturbances, v_u^2 and v_y^2 in the second experiment.

The same formulation process has been used to develop this online version of the IFT algorithm as the standard IFT in Chapter 5 and so all the justification for convergence, stability and robustness for the standard IFT hold true with this online IFT algorithm.

The substitution of the reference for the gradient experiment can be used for enabling online tuning because if the system exhibits a reasonable tracking performance then the new reference for the gradient experiment, $y^1(\rho)$, should be equal to r^1, with some small tracking error. This may also be thought of as the output $y^1(\rho)$ will typically be a low pass filtered version of the reference r^1 and so these signals will be almost identical for a good control system, hence using $y^1(\rho)$ for r^2 will not make the output of the gradient experiment vary far from the normal experiment making this algorithm useful for online applications.

Now after this modification, the output from the first and second experiments are used to calculate $\frac{\partial y(\rho)}{\partial \rho}$ directly in equation (7.3), and therefore a third experiment must be undertaken, using the same conditions as the first experiment to obtain the signal $\tilde{y}_t(\rho)$, so the two signals required to calculate $\frac{\partial J(\rho)}{\partial \rho}$, $\tilde{y}_t(\rho)$ and $\frac{\partial y(\rho)}{\partial \rho}$ in equation (5.7) are unbiased from each other.

Fig. 7.7 gives a schematic view of the new online system implementation. A switch 'S' is used to switch the reference input to the system. In the first and third experiment the switch is set to (1) and for the second, gradient experiment, the switch is set to (11) which feedback the output of the first experiment as there is an N sample delay.

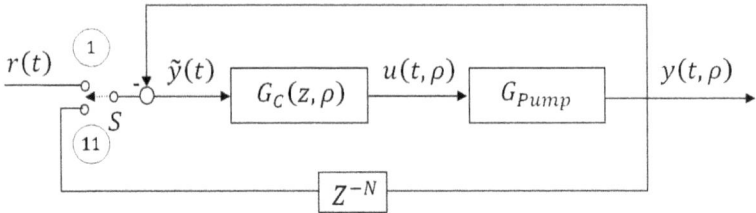

Fig. 7.7. Control system for the online IFT.

An issue may arise when this is implemented if the system is under damped, then the reference for the gradient experiment, $y^1(\rho)$, will contain the exact resonance frequencies of the system, which may cause the system to become heavily excited during this gradient experiment [167]. As explained earlier in the previous section a modified Ziegler-Nichols model-based tuning has been used to ensure the damping is large enough to prevent any unwanted oscillations.

7.2.3. Tuning Comparison in Simulation

The model based and IFT tuning methods that have been developed for the IPMC microfluidic system are implemented on the IPMC model in simulation to validate their performance before implementation. As the IPMC model used is time invariant, this will prove that the IFT algorithm does improve the system performance in an ideal environment. The simulation results for a 0.1 Hz, 100 μm sinusoid reference input are shown in Fig. 7.8. The cost function for the Ziegler-Nichols and IFT tuned systems are shown in Fig. 7.9. It can be seen that both systems start with the same cost, because the IFT algorithm starts tuning from the Ziegler-Nichols based controller parameters, then over the entire experiment the IFT system tunes towards and optimal state whilst the Ziegler-Nichols tuned system remains constant, after the initial transients have passed.

Note the gradient experiment in the IFT tuned system, bottom of Fig. 7.8 which is every third time period, has a slightly larger overshoot than the neighboring experiments. This is due to the fact that the reference for this experiment r^2 is slightly higher than the normal experiments as the system output for the normal experiment y^1, which is used for r^2, does overshoot slightly.

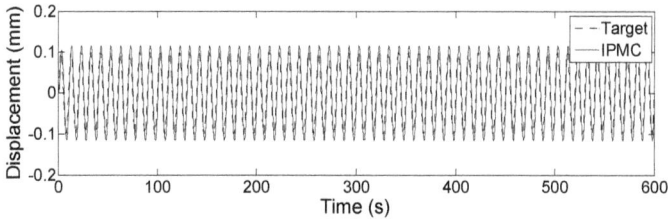

(a)

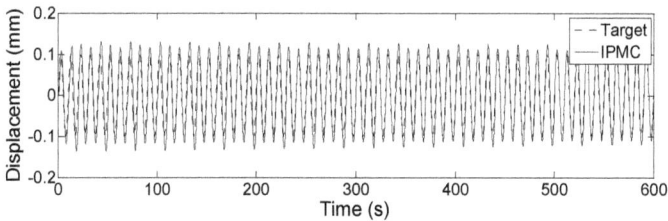

(b)

Fig. 7.8. Simulation of a 0.1 Hz, 100 μm sine wave target signal. Top, Ziegler-Nichols tuned and bottom, online IFT tuned.

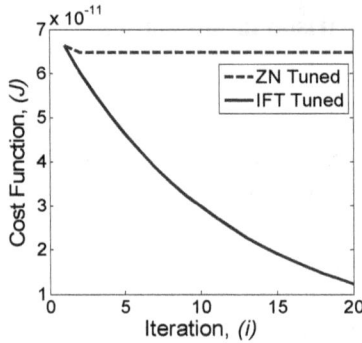

Fig. 7.9. Cost function for the simulated Ziegler-Nichols and IFT tuned system.

7.2.4. Experimental Tuning

7.2.4.1. Experimental Setup

The 17.5 mm long, 10 mm wide, 200 μm thick Nafion® based IPMC, with platinum electrodes was used for all experiments. The test rig shown in Fig. 5.4 was used for the experiments. It accommodates the IPMC in cantilever configuration and allowed voltage to be applied at the 5 mm clamped section. A laser sensor, Banner LG10A65PU, with a 3 μm resolution was used to measure the displacement of the IPMC through a glass window in the Perspex tank which allowed the IPMC to remain hydrated throughout all experiments, replicating the same conditions as in the micropump design. Control electronics and a National Instruments DAQ card were used to interface between the Matlab environment running the Simulink model on the PC and the IPMC actuator. The sampling frequency was 100 Hz and the controller was running at 10 Hz. The step sizes for the IFT algorithm were chosen as 0.05 for K_p and K_i and 0.005 for K_d. The IFT starts tuning from the Ziegler-Nichols model based controller parameters that were found in the model based tuning section 7.2.2.1.

7.2.4.2. Results

A number of experiments were undertaken on the IPMC to test the performance of the control system and online IFT tuning algorithm. Four different sine wave signals were used as a reference input; low

138

and high frequency, 0.05 Hz and 0.1 Hz respectively with low and high amplitude, 100 μm and 300 μm respectively. First, the Ziegler-Nichols tuned controller was implemented on the IPMC, to serve as a benchmark controller and then the IFT algorithm was implemented, which started from the same initial control parameters as the Ziegler-Nichols method. This allowed a direct comparison of a model-based controller with static gains and an online adaptive controller, with dynamic gains. The tracking results are presented below for the four cases as well as the cost function and change in controller gains for the IFT algorithm. Each case was run for 20 iterations which correspond to 20 minutes for the 0.05 Hz signal and 10 minutes for the 0.1 Hz signal.

7.2.4.2.1. Frequency 0.1 Hz, Amplitude 100 μm

It can be seen that the Ziegler-Nichols tuned system exhibits a large overshoot on both sides of operation for the 0.1 Hz at 100 μm targets as shown in Fig. 7.10. This is due to the high frequency of operation, which may be approaching the resonance of the thin IPMC strip. The overshoot may also be caused by the relatively small micro displacement level requiring very precise control. The IFT algorithm also exhibits overshoot at the beginning of the experiment as shown in Fig. 7.10 but then the IFT does improve the system response where the performance of the static Ziegler-Nichols tuned system does not improve. The IFT system improves the design criteria by 82 % where the Ziegler-Nichols actually decreases performance by 22 % as per Fig. 7.11 (a). Overall the IFT ends up 92 % better than the Ziegler-Nichols tuned system. The change in controller gains by the IFT can viewed in Fig. 7.11(b), where it is clear that the final controller is actually quite different from the initial controller showing that the IFT algorithm is doing quite a lot of work.

7.2.4.2.2. Frequency 0.1 Hz, Amplitude 300 μm

The results for 0.1 Hz at 300 μm are presented in Fig. 7.12. The Ziegler-Nichols tuned system does not track the target very well. It can be seen that the IPMC is performing differently on either side of operation. The IFT experiment also starts off with some large overshoot on the same side, but then the gains are quickly adapted as seen in Fig. 7.12 (b), and this causes the system to settle to an acceptable level quickly. Fig. 7.13 (a) shows the continual improvement using the IFT tuned system when compared to the model-based Ziegler-Nichols tuning. The IFT improves by 85 % while the Ziegler-Nichols decreases by 30 %, after 10 minutes the IFT system ends up 89 % better than the Ziegler-Nichols system.

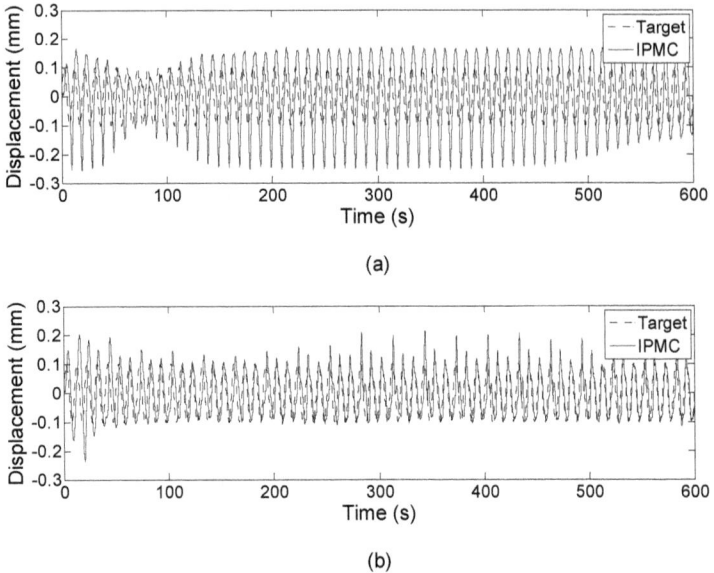

(a)

(b)

Fig. 7.10. 0.1 Hz, 100 μm sine wave target signal.
Top, Ziegler-Nichols tuned and bottom, online IFT tuned.

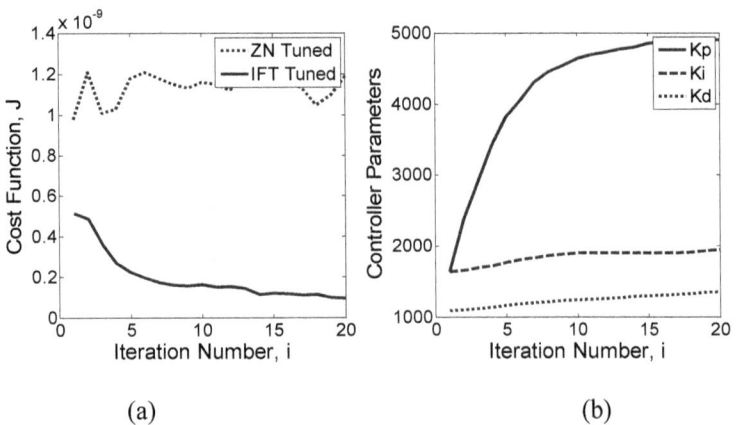

(a)   (b)

Fig. 7.11. (a) Cost function for the Ziegler-Nichols and IFT tuned system and (b) online gain adaptation using IFT algorithm.

(a)

(b)

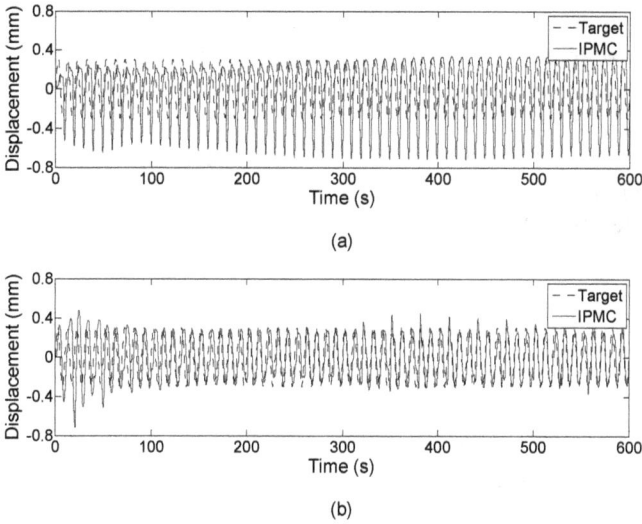

Fig. 7.12. 0.1 Hz, 300 µm sine wave target signal.
Top, Ziegler-Nichols tuned and bottom online IFT tuned.

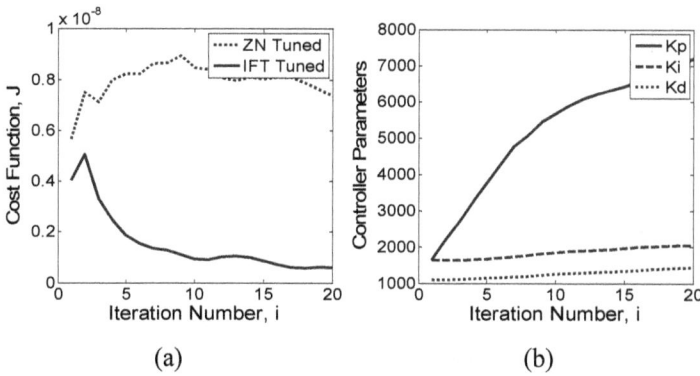

(a)   (b)

Fig. 7.13. (a) Cost function for the Ziegler-Nichols and IFT tuned system
and (b) online gain adaptation using IFT algorithm.

7.2.4.2.3. Frequency 0.05 Hz, Amplitude 100 µm

Fig. 7.14 shows the Ziegler-Nichols system performing fairly well, but does have some overshoot through the entire 20 minute experiment. The performance is better than the 0.1 Hz experiments with the same amplitude, as the IPMC has longer to settle and follow the target, also as the IPMC is further from its resonance frequency. In Fig. 7.14 it can

141

be seen that at the start the IFT algorithm also overshoots, but then the gains are adapted, Fig. 7.14 (b) and much of the overshoot is reduced. This is also visible in Fig. 7.15 (a) in which it is clear that the cost function is decreasing in the IFT tuned system in comparison with the Ziegler-Nichols system. The IFT improves performance by 71 % from the original while the Ziegler-Nichols only improves 15 %. After 20 minutes the IFT system is 66 % better than the Ziegler-Nichols system.

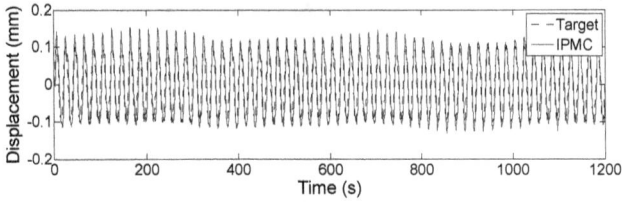

(a)

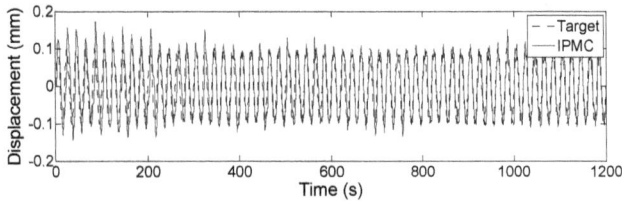

(b)

Fig. 7.14. 0.05 Hz, 100μm sine wave target signal. Top, Ziegler-Nichols tuned and bottom, online IFT tuned.

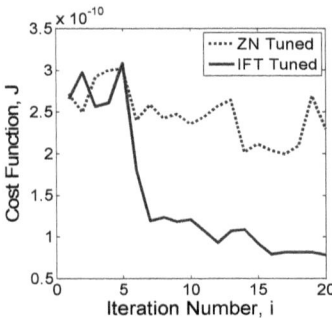 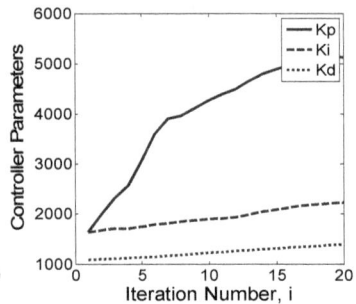

(a)   (b)

Fig. 7.15. (a) Cost function for the Ziegler-Nichols and IFT tuned system and (b) online gain adaptation using IFT algorithm.

7.2.4.2.4. Frequency 0.05 Hz, Amplitude 300 μm

Fig. 7.16 shows the Ziegler-Nichols tuned system where again it can be seen that the system overshoots on one side throughout the entire experiment. This overshoot is also observed at the start of the IFT experiment in Fig. 7.16, but then the system quickly adapts and the system overshoot reduces to a low level. The cost function is given in Fig. 7.17(a) where there is a clear improvement of the IFT algorithm as well as in the conventional model-based approach. As the controller gains for the Ziegler-Nichols approach are static, this improvement is coincidental. The adaptive gains are given in Fig. 7.17(b) where it is clear that the system is automatically tuning itself towards an optimal level, successfully tackling the time varying nature of the system. The IFT approach improves the system performance by 75 % from the first iteration, while the Ziegler-Nichols improves 61 % from its original state. At the end, the IFT tuned system is 83 % better than the Ziegler-Nichols tuned system.

7.3. Experiments with Pump

It has been demonstrated that the IFT algorithm can work over a number of different frequencies and target displacements and so now the IPMC and pump are assembled for testing on the real system. The IPMC is attached to the diaphragm and will push against the diaphragm to pump the fluid. The laser sensor has been positioned so that it can measure the diaphragm displacement. Fig. 7.18 shows the comparison of the diaphragm's displacement with open-loop control, PID control and PID control with the on-line IFT for a targeted diaphragm displacement of 25 μm at 0.1 Hz. From this it can be seen that open-loop control cannot accurately track a desired displacement and the displacement drifts over time which will cause the fluid flow rate to be inconsistent. The standard PID control does maintain the overall mean displacement and prevents the drift that is seen in the open-loop control. The PID controller however is still not very accurate throughout the operation as the controller parameters are not optimally tuned. The PID controller with the new online IFT algorithm is able to control the diaphragm's displacement more consistently over time than a standard PID controller. It does overshoot at the start, similarly to the standard PID controller, but it can adjust its gains over the period of operation to maintain a more accurate displacement and hence flow rate.

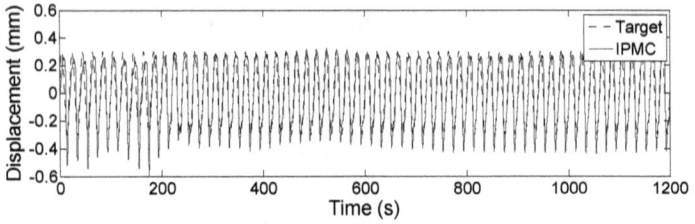

(a)

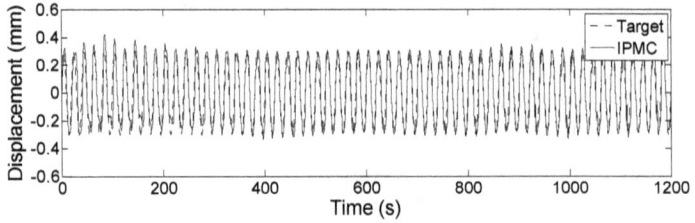

(b)

Fig. 7.16. 0.05 Hz, 300 µm sine wave target signal.
Top, Ziegler-Nichols tuned and bottom online IFT tuned.

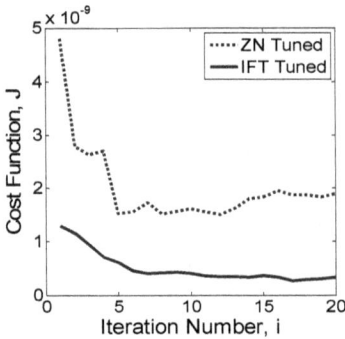

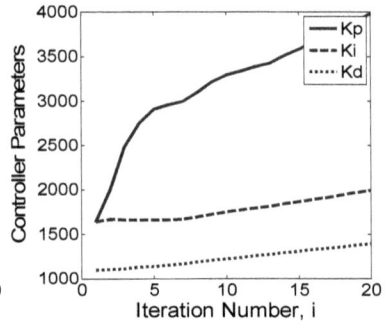

(a)   (b)

Fig. 7.17. (a) Cost function for the Ziegler-Nichols and IFT tuned system
and (b) online gain adaptation using IFT algorithm.

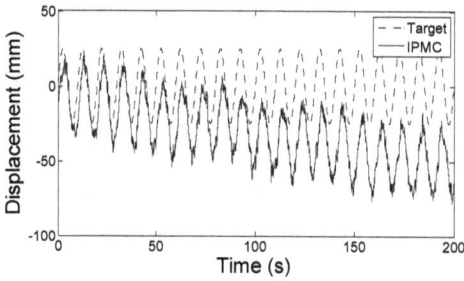

(a)

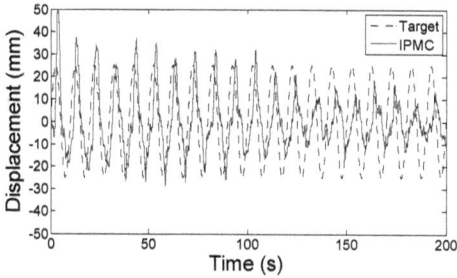

(b)

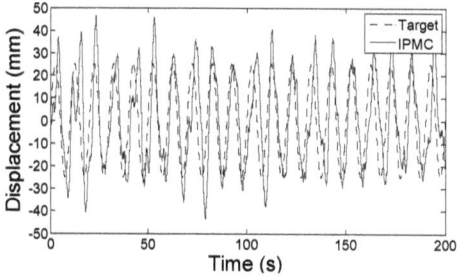

(c)

Fig. 7.18. The displacement of the micropump with (a) OL control, (b) PID control and (c) PID control with on-line IFT [191].

7.4. Discussion

The proposed micropump is simple to fabricate and the design ensures the IPMC is constantly hydrated for extended operation. This means that the proposed device can be used for long periods of time in embedded applications, as the IPMC uses low voltages and can

145

therefore operate for a long time. The main barrier for the IPMC pump to be used for prolonged periods is the reliability of the IPMC performance, as it is inherently time-varying.

In order for IPMCs to be used in the micropump and other real life applications their performance must be guaranteed over a long period of time. This indicates that adaptive control schemes must be implemented to handle the highly time-varying nature of the IPMC. To the authors knowledge no such performance has been tested before in IPMCs, where most experiments are of a simple step response or over some truncated period of time. Even papers which claim to be adaptive [39, 40] do only operate up to 4 minutes. IPMCs have been shown to vary over long periods of time and therefore to demonstrate the success of an adaptive controller the IPMC must be run for a longer period of time. In this research the performance of the IPMC is measured over a long period of operation, up to 20 minutes to show that indeed the IFT system can handle this time-varying IPMC. It is believed and can be extrapolated from Fig. 7.11(a), Fig. 7.13(a), Fig. 7.15(a) and Fig. 7.17(a) that the longer the IFT system runs the better the performance. Therefore it is believed that the IFT system is indeed completely adaptive over the life of the IPMC.

IPMCs have been shown to be hard to control at higher frequencies >0.1 Hz [19] especially as they approach their resonance frequency which tends to be quite low, even less than 1 Hz. It can be seen that the developed control system in this paper can successfully control at 0.1 Hz but the performance is worse than at 0.05 Hz.

The IFT tuning has a modified experiment every third time period, which can cause some deviation from the desired reference. To minimize the effect of this the tuning experiment may be undertaken say every 10th or even 100th iteration, depending on how quickly the system is varying and how quickly the system needs to converge to an optimal state. For example if a system is operating all day, it may need to be tuned for an hour per day to keep the system operating within the required limits.

For the micropump application a periodic signal is used as the input, this is so that the IPMC is always at the origin at the beginning of a new experiment. This is not mandatory as a set point reference could be used or any other reference as long as the three experiments for each tuning iteration are the same. The time period for the experiments can also be changed and as a result the time for tuning will change. For example in the 0.1 Hz case the tuning iteration time is 30 s and in the 0.05 Hz the tuning time is 60 s. If more information is required more than one cycle could be used per experiment in the tuning of the system.

In three of the four experimental cases the IFT has started at a better performance than the model-based tuned system. This is simply a coincidence which depends on the nature of the IPMC and reinforces the fact that even with the same system parameters the performance can be quite different. It has been shown in simulation with a time invariant model that the IFT algorithm does indeed improve the system performance towards an optimal state, Fig. 7.9. This demonstrates that the increase in performance shown in the experimental results can be attributed to the IFT and is not just a coincidence due to the unrepeatable nature of the IPMC. The IFT algorithm has therefore been verified to improve performance in an ideal environment (simulation) as well as a real life noisy environment with external disturbances.

Despite the complex, time-varying and nonlinear characteristics in all cases the IFT algorithm has demonstrated that it does indeed improve the system performance. Also it can be seen that in all cases, after 20 iterations, the system seems to be converging to an optimal solution.

7.5.   Micropump Summary

A micropump driven by an IPMC actuator has been developed. The experimental characterization results show that the IPMC actuator is more effective at low frequencies as it has more time to reach its peak displacement than at high frequency. A pump rate of 130 µl/min at 0.1 Hz has been demonstrated. To guarantee the performance of the IPMC over a long period, the IPMC is constantly hydrated and a PID controller with a new online IFT scheme was used.

The online automatic tuning system has been developed and implemented to adaptively tune the performance of an IPMC which can be used as the active element in the newly developed micropump. The system has successfully tuned the system performance to an improvement of 82 % and 85 % for 100 µm and 300 µm at 0.1 Hz and 71 % and 75 % for 100 µm and 300 µm at 0.05 Hz, respectively.

The system shows that the performance is a lot better with the online IFT tuning than a Ziegler-Nichols model-based tuning method. This shows that a model-free adaptive tuning can outperform a model-based design. This also proves that it is not necessary to spend much time and money to develop an accurate model of the system as if the starting controller is stable then this model free approach can successfully tune the system very well.

Chapter 8

Cell Microtool/gripper and Micromanipulator with Precise and Robust Control

Rapid advances in bioscience and medical research are placing more emphasis on single cell manipulation, for applications such as DNA injection and single cell cloning [192] which have the potential to have a large impact on a variety of diseases such as cancer, diabetes and neuro-degeneration as well as diseases of the cardiovascular system, lungs, blood and skeleton. To enable this research scientists and doctors rely heavily on state of the art tools and equipment to facilitate safe, reliable and precise handling of cells. As a result the demand for novel devices with extensive capabilities to allow researchers to interact with the extremely sensitive and delicate biological materials at the micro and nano scale is becoming huge as these are essential to facilitate this ground breaking medical research [192- 194]. Engineers are therefore being required to play a huge role in developing new intelligent micro and nano manipulators that are more accurate and dexterous than ever before, in order to push the frontier of medical research.

Many current techniques of cell manipulation and handling such as micropipette probing, laser trapping, piezo-grippers, micro-tweezers etc commonly damage cell walls and membranes making them inadequate for many tasks involving 'soft' biological materials like human cells [192, 193, 195]. This has a major effect on the progression of scientific knowledge resulting in significant loss of research time, money and resources in addition to raising ethical issues [194]. To solve these problems pioneering research is required to develop safer manipulation techniques which have more compliance to avoid major cell damage, while maintaining their precision; this is the extremely difficult task which is tackled in this research.

In this chapter a complete novel micromanipulation system is designed using IPMC actuators as a solution to overcome some of the shortcomings of many current manipulation devices by providing a 'soft touch' when dealing with precise micromanipulation tasks and hence enabling safe and reliable manipulation of micro biological materials.

The operating environment during cell manipulation, as well as the physical properties of the cells themselves is likely to be unknown or at least have a large degree of uncertainty and as such robust control methods and algorithms are developed and implemented to control the manipulation system. This chapter presents the development of a new control scheme which is adaptively tuned to optimize both the disturbance rejection (DR) of the IPMC operating in an unknown environment as well as the set point (SP) tracking in order for the manipulator to be both robust and precise. There are very little studies in literature which tackle the DR in IPMCs, some robust controllers have been proposed [6, 79, 113, 116], but this is the first research which directly adapts the system to optimize the response in the presence of unknown external mechanical disturbances. This is one of the most important issues when developing a system which will operate in the real world where the conditions are unknown.

A model-free IFT algorithm has been specifically tailored to adaptively tune both the DR and SP of the system. This is accomplished by introducing a 2DOF control structure which will allow optimal performance under both measures where a traditional 1DOF controller has to either be tuned for the DR or SP and cannot achieve an overall optimal objective [196]. Also by developing an adaptive tuning system the controller will handle the drift in IPMC dynamics and will prove that this system can indeed cope with the time-varying characteristics and hence operate for a prolonged period of time.

The entire integrated system is designed through a large number of simulations before implementing the system and verifying its operation through various experiments to validate the design. These results are presented in this chapter.

The overall contribution of this research is to develop a robust and precise control strategy which adaptively optimizes the DR and SP performance of an IPMC micromanipulator in an unknown environment to achieve accurate position control for undertaking delicate micromanipulation tasks. The manipulator developed in this research demonstrates that it has the ability to further the state of art in cell manipulation and hence be a pioneering system leading the way forward towards truly safe cell manipulation, which will empower further research work by scientists and doctors. The research has been published in [44, 219, 220].

8.1. Single Cell Micromanipulation

To gain more insight and understanding of biological systems, it is necessary to analyze properties of individual cells rather than averaged properties over a population [146]. To achieve this, very complex tasks such as gripping, moving, positioning, altering, controlling and injecting of materials must be undertaken with a high level of precision. This process at the micrometer scale is known as micromanipulation.

At the micrometer scale the effect of adhesion forces is much greater than gravity and so environmental factors (for example relative humidity, temperature) and material properties (for example surface roughness etc.) will greatly affect the reliability and precision of micromanipulation. Safe and reliable manipulation of single biological cells is therefore an extremely ambitious goal to achieve, mainly due to the extremely sensitive nature of cells as well as the complex biological operating environment which tends to be unknown, especially at the micro and nano level. As such it is highly desirable to have a micromanipulation technique which includes system compliance, and hence safety.

The required manipulator specifications are therefore high precision control with a guaranteed level of compliance to give a sensitive touch to avoid damage to the specimen, as well as guaranteed DR and adaptation to compensate for all unknowns. Current micromanipulators include micro-grippers, micro-needles, micro-injectors, micro-tweezers, micro-tubes, micropipettes, micro-cantilevers and so forth [193]. Traditionally all these devices have the downfall of high stiffness which can cause issues when operating in highly unknown environments. Some examples of current techniques are described below.

8.1.1. Current Techniques

Various solutions for micro-gripping for MEMS assemblies have been proposed [197, 198], but only a few are suitable for bio-micromanipulation [146, 199]. Most existing bio-micromanipulation techniques are non-contact, like laser trapping [200, 201] and electro-rotation [202, 203]. Laser trapping uses high energy laser beams and can cause localized heating and hence damage to the cell while electro-rotation cannot be used to keep a cell in the same fixed position.

Current contact micromanipulation techniques typically involve two conventional manipulators and micropipettes with 3DOF motion, schematic shown in Fig. 8.1, which is highly geared permitting a high resolution but consequently results in very high stiffness and no back-drivability or compliance. These systems typically operate in open-loop with the feedback from a microscope being processed by the user and manual control is needed. Manipulator operations, especially cell rotation, are extremely difficult for inexperienced users and typically it takes more than 6 months of training to become proficient at operating a system [192]. This method also relies on bulky and expensive setups.

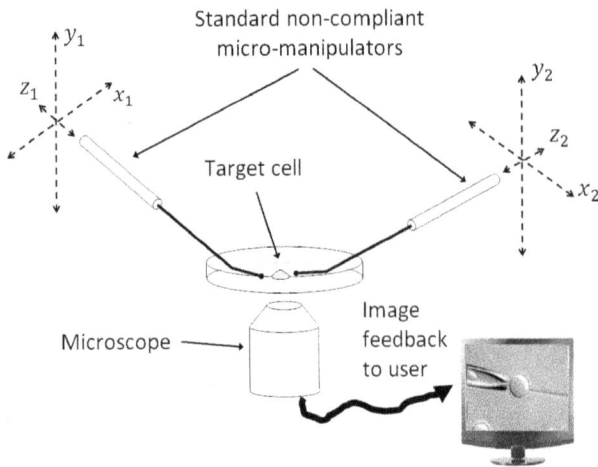

Fig. 8.1. Conventional cell micromanipulation system.

An improvement on this standard manipulator system has been proposed by Inoue *et al.* in [192], which consists of a 3DOF parallel mechanism to control the worktable and a two fingered 'chopstick' like probe mechanism. While this gives improved resolution the stiffness of the system is increased further as the actuators act in parallel. A soft handling probe was developed in 2010 by Takeuchi *et al.* [204] to give some compliance through using a thermal gel to manipulate cells. However it was found that the size of the generated thermal gel was too large to manipulate micro objects and further work is needed. A bio cell processor for single cell manipulation using a polypyrrole valve, has been designed to handle individual embryo cells in [205], however this still does not meet the current requirements due to lack of dexterity and range of tasks it can perform, hence further improvements are needed.

Compliance can be actively implemented on standard devices, typically implemented through complex back-drivable algorithms and impedance control with expensive force sensor feedback. This alone may often be unreliable in unknown micro and nano environments where electronic and mechanical-thermal noises are commonly large and the control system may not react in a timely fashion to avoid damage. An actively compliant probing system has been developed in [206] but this is designed for in-circuit testing of PCBs and as such is not suitable for cell manipulation.

Typically high precision comes at the cost of high stiffness. IPMCs circumvent this issue through their natural passive compliance adding inherent compliance characteristics without sacrificing any accuracy in tracking performance. IPMCs can be used to couple high precision with high compliance to create a safe and accurate micromanipulation system.

8.1.2. IPMCs for Micromanipulation

IPMCs have intrinsic properties which make them desirable sensors and actuators for micromanipulation, including very low mass, custom geometries, biocompatibility, flexibility and compliance, low power consumption and the ability to accurately achieve both micro and macro deflections without any gearing or other complex mechanisms [5,19]. IPMCs present a unique and smart system for cell manipulation due to their ability to work well in fluid and cellular environments as well as their natural compliance giving them a 'soft touch' when interacting with sensitive materials and/or in sensitive environments.

IPMCs natural compliance makes them excellent candidates for safely handling sensitive biological materials. The passive compliance of a beam type IPMC manipulator, which is the type concentrated on in this research, is defined as

$$S_{manip} = \frac{1}{k_{manip}} = \frac{L^3}{3EI}, \qquad (8.1)$$

where s_{manip} is the mechanical compliance and k_{manip} is the stiffness of the manipulator. The compliance depends on the IPMC properties, free length, L, modulus of elasticity, E and moment of inertia, I. This is very useful in designing a manipulator for a specific application as the IPMC dimensions and material properties can be tailored to meet the manipulator requirements for force, displacement and compliance. Also as IPMC have electrically induced displacement IPMCs can be controlled to have a variable compliance through implementation of an active compliance controller. This requires force feedback which is outside the scope of this research. Passively compliant IPMC manipulators, as developed in this research, do have advantages over an actively controlled compliant system including guaranteed compliance throughout operation, no need for complex control algorithms, safety achieved without loss of positioning performance and no expensive force/torque feedback sensors are needed.

On the other hand IPMCs also exhibit inherent nonlinearities and time-varying behaviors which have been the major stumbling block to implementing them into real life applications as replacements for traditional sensors and actuators. To overcome this, in order to carry out useful micromanipulation tasks, robust and intelligent design must be coupled with the IPMCs to enhance their capabilities and add the high level of performance and reliability which is necessary to achieve their goals.

IPMCs are showing major promise as cell manipulators, because of their inherent compliance and the recent advances in IPMC control they have the unique ability to act as precise and reliable manipulation systems which are superior to current devices [192, 196, 204].

Some microgripper devices which use IPMCs for micromanipulation have been proposed, for example in [146, 207] but no closed-loop control is implemented. In [208] a simple linear position control is implemented on three separate IPMCs of a three finger gripper to simulate a pick up and drop motion, but no consideration has been given to the response when an external force is applied due to the object being grasped. An integrated IPMC/PVDF actuator/sensor has been successfully used for the open-loop micro-injection of living Drosophila embryos [209].

The micromanipulation system in this research builds on the previous IPMC devices to develop a more precise, robust and complete micromanipulation and positioning system. The closed loop controller also accounts for the effect of the disturbances due to the load being manipulated.

8.2. Micromanipulator System Design

8.2.1. Microtool/gripper Design

IPMC actuator motion, with appropriate control, is well suited to the precise and delicate manipulation tasks which need to be undertaken by medical researchers, such as rotating, gripping and placing cells, probing or injecting cells, applying forces and deforming cells as well as general moving of cells in their environment. In order to achieve these tasks IPMCs can be manufactured into custom sizes and geometries required for a specific micromanipulation task, and they can easily be scaled without losing their sensing and actuation properties. Also any number of different types of microtool can easily be attached to the end of an IPMC in order to give the manipulator the desired functionality. In this research a beam shaped IPMC will be used as a microtool/gripper as this shape is very versatile for a number of different configurations and tasks, as shown below in Fig. 8.2. Also the model developed in Chapter 3 is based on a beam shaped actuator and so this can easily be used to simulate the performance of the manipulator designs.

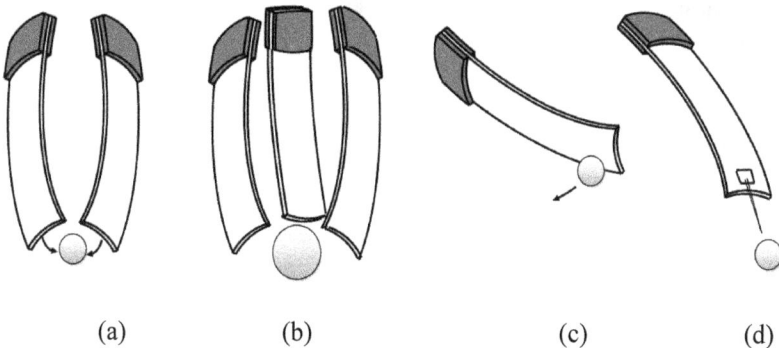

(a)	(b)	(c)	(d)

Fig. 8.2. Precise and sensitive micromanipulation tasks for IPMCs. (a) Two IPMC gripper, (b) three IPMC gripper for more difficult to handle or larger payloads, (c) pushing/pulling and (d) probing or injecting cells.

8.2.2. Micromanipulator Design

8.2.2.1. Specifications for Proposed Manipulator

In order to achieve safe and reliable manipulation of highly sensitive biological cells in a complex biological operating environment, which tends to be unknown, the required manipulator specifications must be high precision control with a guaranteed level of compliance to give a sensitive touch to avoid damage to the specimen. On top of this the system must be able to adapt to the changing environment, changing tasks for example loaded/unloaded and drift in the dynamics of the IPMC itself.

For the proposed system to be useful for operating in real experiments, a number of common tasks must be achievable, namely gripping, pushing/pulling, probing and injecting cells. To realize this, a number of crucial specifications are required for the system, all of which have been considered throughout the design process. These specifications, outlined below, are based on current devices and literature for the manipulation of biological cells [193, 196, 210]. The specifications are split into two groups. The first six specifications will be met by the mechanical system which is developed using a thorough design process, including modeling and simulations:

1. High dexterity (DOF) to achieve required tasks;

2. High compliance and hence back-drivability to achieve a 'soft touch';

3. < 10 μm resolution;

4. > 4 × 4 mm ROM;

5. Ability to accommodate specific microtools;

6. Consideration for all sensors and a microscope.

The second groups of specifications, 7-9 below, are met through the development of an advanced control system. The required specifications for the control system are to,

7. Optimize the DR;

8. Optimize the SP tracking response, and

9. Adapt to the time-varying system dynamics,

in order to achieve high accuracy and repeatability for all manipulation tasks. The following section describes how specifications 1-6 have been met.

8.2.2.2. Mechanical Mechanism Design

In order to develop such a system with adequate level of dexterity to perform all the required tasks it was decided to utilize a modular system design. A standard module is designed using two IPMC actuators to give 2DOF motion. This design keeps the modules simple and robust, with only two rotary joints, making them easy to implement and with minimal friction, stick-slip behavior, backlash and other mechanical loses and inaccuracies. Each IPMC must be clamped by 5 mm at one end in order to provide electrical stimulus while free to bend at the other. The force provided by the IPMC due to the input electric potential must then be transferred to the manipulator mechanism which will interact with the sample specimen through the attachment of an appropriate microtool.

A number of different concepts were considered based on the above criteria and specifications, each design was iterated through until a final design was developed. An individual module of the final design is presented in Fig. 8.3. It can be seen that there are two rotating arms which convert the non-uniform bending of the IPMCs into a uniform rotational path by using the 'tip slots' to allow the IPMC tip to freely slide in. This mechanism gives a more consistent and hence robust motion while adding minimal frictional losses and backlash to the system.

IPMC #1 drives the top arm (red arm) to produce the horizontal motion, θ_1, while the bottom arm (blue arm) rotates due to IPMC #2 to give vertical motion, θ_2. The top arm is connected to the base through 'Pin joint #1', the bottom arm is connected to the top arm though 'Pin joint #2' and a microtool is fixed to the bottom arm. The microtool end effector motion is therefore the result of the horizontal and vertical components of motion, giving a 2DOF ball joint type actuation. The length of microtool end effector can be adjusted to modify the range and resolution of the system; this is governed by the lever principle- the longer the end effector, the larger the range and the smaller the resolution, and vice versa. Also the longer the arm, the smaller the available force to apply to the target cell, and vice versa. In this way a number of different tasks can be achieved depending on the size of cell and required force etc.

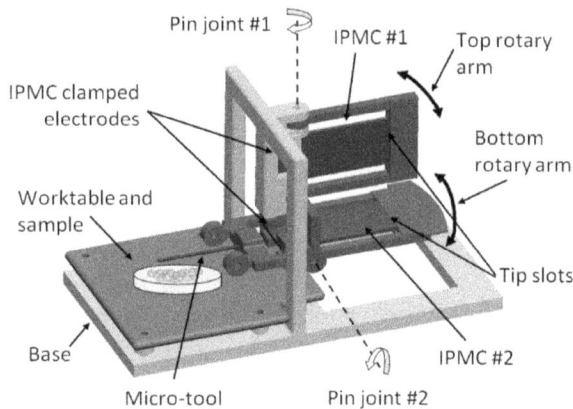

Fig. 8.3. Individual 2DOF micromanipulation module [219].

IPMC #1 acts horizontally and hence does not have to overcome gravity, while the arm for IPMC #2 has been counter balanced to ensure that despite acting in the vertical direction the gravitational effect will be minimal.

In Fig. 8.4 (a) and (b) a conceptual drawing is given of the possible layout of two and three modules of the IPMC micromanipulators. By adjusting the angles and orientations of the modules any number of tasks can be undertaken by the system. Through this innovative modular design, the system can be controlled simply, yet the overall dexterity of the manipulator can become extremely high with a large number of DOF possible.

Two laser displacement sensors, with a 10 μm resolution, will be used to provide displacement feedback for the 2DOF, as shown in Fig. 8.5 (a) and (b), in order to control each modules response. A microscope can easily be placed above or under the transparent worktable in order to provide visual feedback for the user as shown in Fig. 8.5 (b). The worktable can also be adjusted up and down to accommodate different biological samples.

(a)

(b)

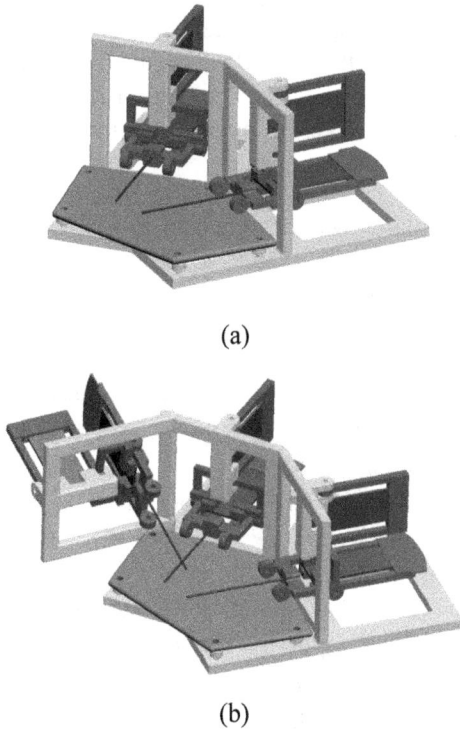

Fig. 8.4. Conceptual design of multi-micromanipulation system with (a) 2 modules and (b) 3 modules [219].

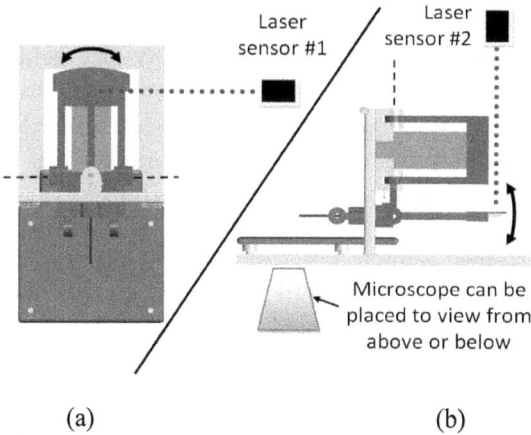

(a)   (b)

Fig. 8.5. (a) Top and (b) side view of a micromanipulation module with sensors and microscope.

The two arms can be designed to whatever dimensions are required to accommodate any particular size IPMCs (as IPMCs can be fabricated to any geometry) for the specific application requirements, i.e. force, displacement etc. This gives a huge flexibility to this manipulator design.

The robotic manipulator was designed in Pro/ENGINEER and then a mathematical model was developed to predict the dynamic response. The mathematical model, in joint space representation, is given in equation (8.2). This has been developed to fully describe all the mechanical dynamics of the mechanism, including inertia, Coriolis effect, gravity and joint friction of the manipulator so accurate simulations can be undertaken.

$$M(\theta)\ddot{\theta} + C(\theta,\dot{\theta}) + G(\theta) = \tau_{IPMC} - \tau_{JointFrict} , \qquad (8.2)$$

where M is the mass matrix, C is the vector of centrifugal and Coriolis torques, G is the vector of gravitational torques. These effects are balanced by the torque input by the IPMC, τ_{IPMC}, minus the joint friction torques, $\tau_{JointFrict}$, $\ddot{\theta}, \dot{\theta}$ and $,\theta$ is the individual joint angular acceleration, velocity and position respectively.

8.2.2.3. IPMC Actuators for the Manipulator

In this design it is important to make full use of the available IPMC materials as two actuators are needed for each module. IPMCs can be fabricated to any geometry and therefore any size actuator could be used for the system depending on the performance requirements. In this project a 55 × 20 mm, 800 μm thick IPMC sheet, from ERI based on an XR resin, with Pt electrodes was available for use. This sheet was the total amount of material to be used to cut IPMC strips from. It was therefore extremely important to have developed the IPMC model so that the mechanical outputs for different size strips can be simulated and an appropriate size of actuator, to move the mechanism the required distance as well as provide the required force on to the target cells, was chosen.

The model developed in Chapter 3, with updated material parameters for the new IPMC material that was to be used in this research, was used for the mechanical design process to simulate the performance with different sized actuators before the system was actually built. The model does accurately represent the IPMC response at the time it was developed, but the IPMC may change response unpredictably. Therefore the simulations will give an indication of the system performance but it can be expected that the experimental results may vary from the simulation results.

8.2.2.4. Manipulator Optimization and Validation

The full manipulator model is now available and any different size manipulator can be simulated as both the mechanical and IPMC model are geometrically scalable. Now depending on the required specifications a number of manipulators can be developed to carry out a number of different tasks. For this research the specifications have been laid out in Section 8.2.2.1 and the system must be designed to meet these with the available IPMC sheet.

With the proposed modular design multiples of two IPMC actuators are needed (2, 4, 6, etc) to be cut from the available IPMC sheet. All the possible configurations for cutting the sheet which were considered are shown below in Table 8.1.

Table 8.1. Possible configurations for cutting sheet of IPMC material [219].

# IPMCs	Configurations for cutting IPMC sheet		
2×	(i)	(ii)	
4×	(iii)	(iv)	(v)
6×	(vi)	(vii)	(viii)

The geometrical details for all configurations are given in Table 8.2, where the aspect ratio equals (width/length):1. All configurations were considered by simulating the performance, using the complete IPMC and manipulator mechanical model, in order to find the best manipulator performance. The mechanical compliance of each IPMC actuator is also given in the table below.

Table 8.2. Possible configurations for manipulator modules
for available IPMC sheet.

Configu -ration	Modules	# IPMCs	Width (mm)	Length (mm)	Aspect ratio	Compliance (m/N)
i	1	2	10	55	0.18 : 1	0.304
ii	1	2	20	27.5	0.73 : 1	0.0138
iii	2	4	5	55	0.091 : 1	0.608
iv	2	4	13.75	20	0.69 : 1	0.00597
v	2	4	10	27.5	0.36 : 1	0.0277
vi	3	6	3.33	55	0.061 : 1	0.913
vii	3	6	9.17	20	0.46 : 1	0.00895
viii	3	6	6.67	27.5	0.24 : 1	0.0415

Both configurations (i) and (ii) will give good performance in both force and displacement, but it was decided that for this particular design at least two modules would be needed to have the required dexterity to perform gripping and injecting tasks. Configurations (iii) and (vi) have the largest displacement output, but will struggle to move the manipulator accurately as they would be flimsy due to their low aspect ratio. (iv) and (vii) have the highest aspect ratio and this gives them a high force output as the force output is highly dependent on the width of the IPMC strip, however they do have limited displacement output due to their short length, only 15 mm of free bending. It was found that (v) and (viii) give the best response with large deflections and sizable force available. Configuration (viii) has a higher compliance than (v) although configuration (v) will have a larger force output, due to a larger width, and very similar displacements to (viii), as the free bending length is the same. Therefore it was decided to use

(v) as the model simulations are only approximate and it was more important to oversize the actuator so that it will have ample power to overcome any mechanical losses that may have been underestimated in the simulation, like the stick-slip phenomena as well as the weight of cooper electrodes and the electrical cable attachments etc. Also having 2 modules will be adequate for all of the tasks which will be required in cell manipulation as in current systems. The results of the simulation of a piece of the IPMC sheet which measures 27.5 mm long by 10 mm wide (5 mm clamped section), chosen for configuration (v), with 1, 2 and 3 V inputs is shown in Fig. 8.6.

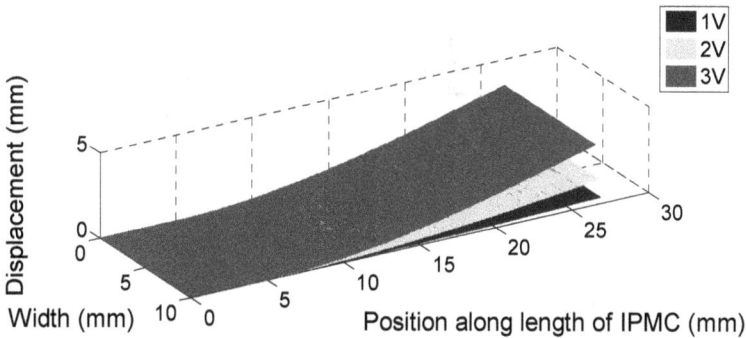

Fig. 8.6. Model simulation of a 27.5 mm long by 10 mm wide IPMC with clamped length of 5 mm, under a 1, 2 and 3 V input [219].

If different amounts of materials were available the same optimization process could be undertaken to decide the configuration or if a particular desired performance is required, the minimum amount of IPMC material required can be calculated before fabricating the IPMCs from the raw materials to avoid any waste.

Open-loop simulations of the manipulator module reveal that up to 8.6° can be achieved in both axes with a saturated 3V input. Fig. 8.7 shows an open-loop simulation of the two axis of motion with an out of phase 3 V pulsed input for configuration (v) cut from the IPMC sheet.

The manipulator motion represents a ball type joint, which can be thought of as the combination of two levers, one for horizontal and other for vertical motion. The resolution and ROM of the microtool are dependent on the distance of the tip of the microtool to the COR of the two arms.

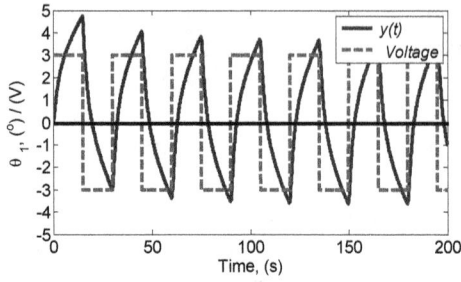

(a)

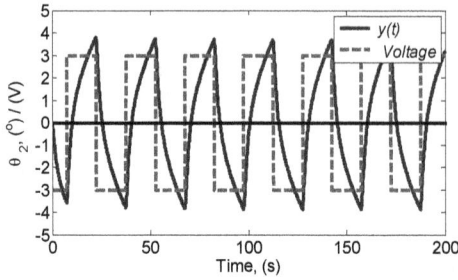

(b)

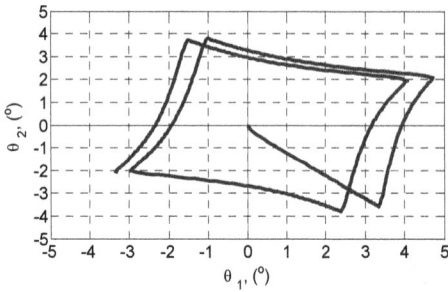

(c)

Fig. 8.7. Open-loop simulation results for (a) horizontal, (b) vertical and (c) end effector path [219].

The resolution of the manipulator end effector depends on the laser sensors (which have a 10 µm resolution) and because specification 3 states the manipulator must have a resolution less than 10 µm, the laser sensors must be placed further away from the COR than the tip of the microtool is (lever principle). The IPMC length has been chosen to be optimal at 27.5 mm (22.5 mm free deflection) and so the laser target will be placed at 30 mm back from the COR. Therefore to meet the resolution specification the microtool tip has to be < 30 mm from the COR.

Specification 4 states the ROM must be greater than 4mm in both axis and from the open-loop simulations it was found that 8.6° can be achieved. With this angle the tip must be >13.38 mm from the COR to meet the ROM specification.

Through this simulation the optimal configuration for cutting the IPMC actuators from the sheet has been chosen and after this a number of different resolutions, ROMs and forces can be achieved by using microtools with different lengths which are between 13.38 mm<L_{COR} <30 mm all of which meet the specifications laid out. In this design a length of tool was chosen as 20 mm as this fits well in between the required specifications.

8.2.3. Complete Manipulation System

The micromanipulator has been designed to meet all its specifications and a number of microtools have been proposed as well. The manipulation system can interact with the target cell itself using a conventional micropipette and it will still be safe due to the compliance of the IPMC actuators. The proposed microtools can be used with a conventional 3DOF positioning system and will possess compliance and safety due to the flexible IPMC actuators. The complete system may be constructed where the IPMC micromanipulator is fitted with an IPMC microtool on the end to interact with the cells. This is a complete IPMC micromanipulation system with two levels of compliance. The first level is the microtool and if that fails the manipulator is also compliant to avoid damage.

8.3. Micromanipulation Control

Theoretical simulations have demonstrated that the system can meet the resolution and ROM specifications but closed loop control is required to ensure that the system can perform accurately and reliably in a complex and unknown environment and when interacting with sensitive biological materials. The system will inevitably be operating in the presence a large amount of electronic and mechanical-thermal noises as well as nonlinearities introduced due to mechanical friction and backlash of the mechanism as well as other external disturbances to the system. Also the IPMC actuators themselves are highly non-linear and time-varying so this presents a major control problem which requires an adaptive control algorithm to be employed.

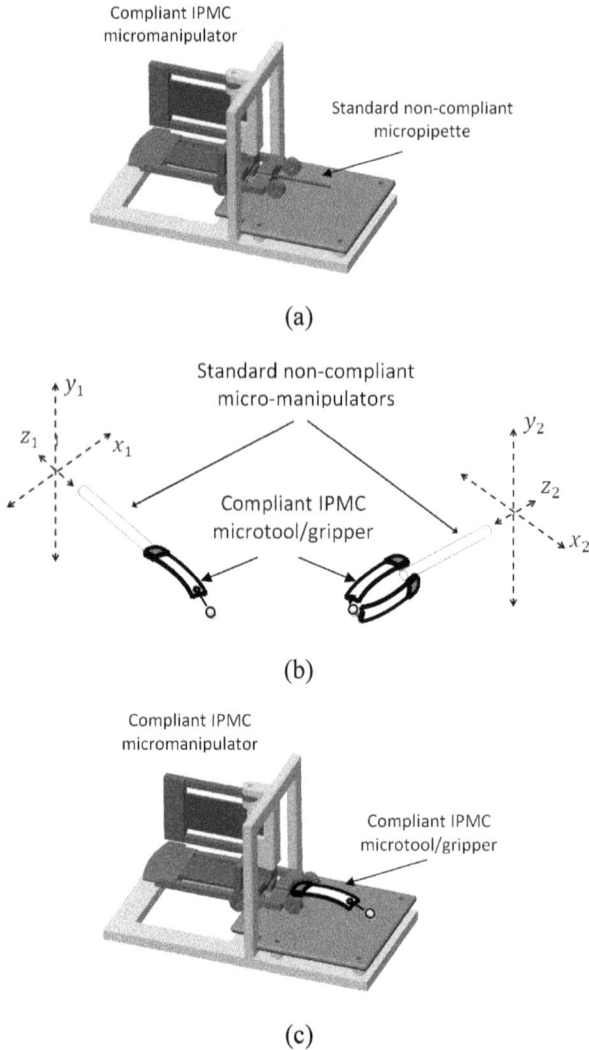

(a)

(b)

(c)

Fig. 8.8. Integrated compliant IPMC micromanipulation systems.
(a) Compliant IPMC manipulator with standard micropipette;
(b) standard non-compliant micromanipulators with compliant IPMC microtool/gripper and (c) compliant IPMC manipulator with compliant IPMC microtool/gripper.

In order to meet the specifications for the controller a number of different control system architectures were considered. A 2DOF control system was then proposed which could be optimally tuned to provide both excellent DR while also achieving precise micro position control where a traditional 1DOF controller has to either be tuned for the DR or SP and cannot achieve an overall optimal objective [196]. In order to achieve the optimal performance and adapt to the changing system dynamics an A specifically tailored IFT algorithm has been proposed to adaptively tune both the DR and SP of the manipulator. Through the innovative combination of these techniques all the required specifications can be met. This new control system will be compared to a standard 1DOF controller to demonstrate its superior performance.

8.3.1. 2DOF Control Structure

A 2DOF control system configuration has two closed loop transfer functions between the input and output which can be adjusted independently [211]. Control system design is a most commonly a multi-objective problem and as such a 2DOF controller has a natural advantage over a standard 1DOF controller with one closed loop transfer function. This 2DOF configuration has been specifically chosen for the micromanipulation application due to the benefit of optimizing both the DR of the output as well as the SP or tracking response of the system. This is exactly what is needed for the IPMCs to robustly operate in an unknown environment.

The 2DOF system which will be used in this system is shown below in Fig. 8.9. It consists of a feed-forward regulator, $G_f\left(\rho_f\right)$ and a serial regulator, $G_c\left(\rho_c\right)$ which are both expressed here in terms of their tuneable parameters ρ_f and ρ_c. The two regulators determine the performance of the manipulator, G_{manip} by controlling the input voltage $u(t)$. H_S represents the laser sensor dynamics and will be combined with a low filter pass filter to reject the majority of the measurement noise in the system, $d_m(t)$. The objective is that the IPMC deflection, $y(t)$ tracks the desired reference, $r(t)$ with minimum error, $\tilde{y}(t)$ and that all external disturbances $d(t)$ are suppressed. This objective is achieved by optimally tuning the two regulators in the system. The design criteria to be minimized and used to measure the performance of the IPMC manipulator is a least squares fit of the tracking error, as shown previously in equation (5.27).

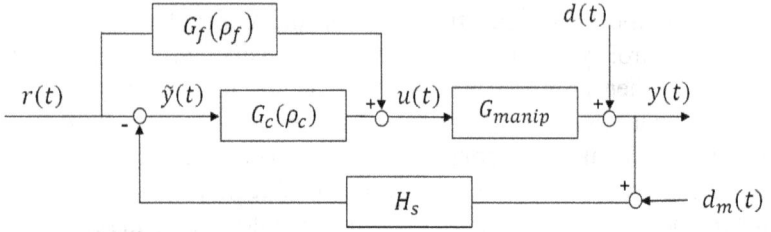

Fig. 8.9. 2DOF controller structure.

The closed loop transfer functions from r to y and d to y for the 2DOF controller are expressed in the Laplace domain as follows,

$$G_{yr2}(s) = \frac{G_{manip}\{G_c(s)+G_f(s)\}}{1+G_{manip}G_c(s)H_s(s)} \qquad (8.3)$$

$$G_{yd2}(s) = \frac{1}{1+G_{manip}G_c(s)H_s(s)} \qquad (8.4)$$

It can be seen from equation (8.4) that the effect the disturbances $d(t)$ have on the output $y(t)$ are fully determined by G_C. The DR can therefore be fully described by G_C and hence is tuned through the parameters ρ_C. From equation (8.3) the SP response is determined by both regulators G_C and G_f. As G_C is tuned to regulate the DR, the SP can be determined by tuning the parameters ρ_f.

8.3.2. Stability

It can be shown that the steady state error to a unit step change of set-point $r(t)$ and the steady state error to a unit step disturbance $d(t)$ will become zero robustly in this system if

$$\lim_{s \to 0} G_c(s) = \infty \qquad (8.5)$$

$$\lim_{s \to 0} \frac{G_f(s)}{G_c(s)} = 0 \qquad (8.6)$$

$$\lim_{s \to 0} H_s(s) = 1 \qquad (8.7)$$

To meet the conditions imposed on the regulators from equations (8.5) and (8.6) it was decided to include an integrating element in G_C and not in G_f. The regulator G_C is now made up of the sum of a proportional, integral and derivative element while the feed-forward regulator has just proportional and derivative elements as shown in equations (8.8) and (8.9). The 2DOF regulator parameters shown in Fig. 8.9 can therefore be described as $\rho_c = \begin{bmatrix} K_p K_i K_d \end{bmatrix}^T$ and $\rho_f = \begin{bmatrix} \alpha \beta \end{bmatrix}^T$.

$$G_c = K_p + \frac{K_i}{s} + K_d s \qquad (8.8)$$

$$G_f = \alpha + \beta s \qquad (8.9)$$

The condition in equation (8.7) requires that the output feedback loop be accurate in steady state. A first order low pass filter of the form in equation (8.10), has been implemented with the sensor readings to ensure this condition holds true. The filter is also used to reject the high frequency measurement noise, $d_m(t)$.

$$F_{lp} = \frac{1}{1 + s\tau} \qquad (8.10)$$

8.3.3. Tuning Procedure

The system described above guarantees steady state stability, now the objective is to tune the regulator parameters to achieve optimal performance. It can be seen from equation (8.4) that the effect any disturbance $d(t)$ has on the output $y(t)$ can be tuned by only adjusting the parameters in G_C. The DR can therefore be fully described by G_C and hence the DR problem can be optimized by tuning the parameters ρ_C. Once these parameters have been found the next step is to optimize the SP and it can be seen from equation (8.3) that this involves both the regulators, although as GC has already been optimized so the SP problem then is to optimally tune ρ_f. This two-step method has advantages in that standard PID tuning techniques can be employed in the first step and that the number of variables to optimize is not large. However this method does not necessarily guarantee the global optimum, if for instance the poles of the system that are determined in the DR tuning are too extreme then only a relatively poor set point response may be possible [196]. In this research the two step tuning has proved to be extremely effective for optimizing both the DR and SP as can be seen in the results sections of this chapter.

8.3.4. IFT for 2DOF Controller

The 2DOF controller must be tuned for two conditions the DR and SP. As explained in the previous section the DR will be tuned first.

In order to solve $\frac{\partial J(\rho)}{\partial \rho}$ at any instant two signals are needed, $\tilde{y}_t(\rho)$ and an estimate of $\frac{\partial J(\rho)}{\partial \rho}$. These must be found independently such that they are unbiased from each other.

A first experiment can be conducted under normal operating conditions, i.e. r set to zero for DR tuning. $\tilde{y}(\rho)$ can then be found from $\tilde{y}(\rho) = r - y^1(\rho)$.

To calculate $\frac{\widehat{\partial y(\rho)}}{\partial \rho}$, for iteration i, the output y for a given control system to be tuned needs to be differentiated with respect to its tuneable parameters. $\frac{\widehat{\partial y(\rho)}}{\partial \rho}$ for the 2DOF DR tuning is calculated from equation (8.11). This is the equivalent of equation (5.14) from the standard IFT algorithm (ignoring disturbances under the IFT assumptions).

$$\frac{\widehat{\partial y}}{\partial \rho}(\rho) = \frac{1}{G_c}(\rho)\frac{\partial G_c}{\partial \rho}(\rho)\left[\frac{G_c(\rho)G_{manip}}{1+G_c(\rho)G_{manip}H_s}(-y(\rho))\right] \quad (8.11)$$

When tuning for DR the reference input is zero and so Gf is redundant. By comparing the term in the square brackets in equations (8.11) and (8.3) it can be seen that the term in the square brackets is the result of injecting the error from the first DR experiment through the closed loop system when $G_f(\rho) = 0$. The output from the plant, y^2, then gives the term in the square brackets in equation (8.11). The two terms $\frac{1}{G_c(\rho)}$

and $\dfrac{\partial G_c(\rho)}{\partial \rho}$ can be calculated from equation (8.8) and hence

$\dfrac{\widehat{\partial y}(\rho)}{\partial \rho}$ can be established and the new improved parameter update for DR is found.

Once the DR parameters, ρ_C, have been tuned the SP parameters, ρ_f, are tuned next. This is done by conducting a first experiment, by setting the reference to a desired set point value which is representative of the references the system will be subjected to during normal operating and calculating $\tilde{y}(\rho)$ from $\tilde{y}(\rho) = r - y^1(\rho)$.

To calculate $\dfrac{\widehat{\partial y}(\rho)}{\partial \rho}$, for SP tuning the control system is differentiated with respect to its tuneable parameters. $\dfrac{\widehat{\partial y}(\rho)}{\partial \rho}$ for the 2DOF SP tuning is calculated from equation (8.12) (ignoring disturbances under the IFT assumptions).

$$\frac{\widehat{\partial y}}{\partial \rho}(\rho) = \frac{1}{G_c(\rho) + G_f(\rho)} \frac{\partial G_f}{\partial \rho}(\rho) \left[\frac{G_{manip}\{G_c(\rho) + G_f(\rho)\}}{1 + G_{manip}G_c(\rho)H_s} r \right] \quad (8.12)$$

Similarly to DR tuning the square brackets in equation (8.12) can be found by injecting the standard SP reference, r, into the system with the 2DOF closed loop controller, equation (8.3). The terms in front of the brackets can be found analytically from equations (8.8) and (8.9). In this way $\dfrac{\widehat{\partial y}(\rho)}{\partial \rho}$ can be established and the new improved parameter update for SP is found.

A point to note is that both the DR and SP can be tuned online through normal operation because in the gradient experiments for both conditions the reference does not deviate far from the reference for the normal experiment. In DR $r^2 = -y^1$ (i.e. the error from the first experiment, under normal conditions) and so if the system is tuned well then r^2 should be close to zero, which is the same reference for the normal experiment. Also for SP tuning $r^2 = r$ and so the gradient and normal experiments are the same. Note two experiments are still needed to ensure $\tilde{y}_t(\rho)$ and $\widehat{\partial y(\rho)}\big/\partial\rho$ are independent from each other. This makes this new tuning method very useful for real-life systems which are meant for continuous operation.

8.3.5. Comparison with 1DOF Controller

To provide a benchmark to measure the new proposed control algorithm against a standard 1DOF PID controller was also implemented on the manipulator. The traditional 1DOF system is shown in Fig. 8.10 and the transfer functions which describe the behavior between the reference and disturbance to the output are given in equations (8.13) and (8.14). It can be seen that both the SP and DR depend on the controller $G(\rho_1)$, hence the 1DOF parameters, ρ_1. As there is only one tunable element for both DR and SP their performance cannot be adjusted independently. This causes difficulty in design as typically if the disturbance response is optimized then the SP is often found to be poor and vice versa [196]. The 1DOF controller will have all proportional, integral and derivative parts and hence $\rho_1 = \begin{bmatrix} K_{p1}K_{i1}K_{d1} \end{bmatrix}^T$.

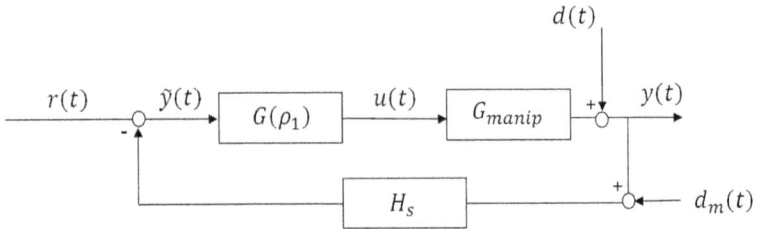

Fig. 8.10. 1DOF control scheme.

$$G_{yr1}(s) = \frac{G_{manip}G(s)}{1 + G_{manip}G(s)H_s(s)} \quad (8.13)$$

$$G_{yd1}(s) = \frac{1}{1 + G_{manip}G(s)H_s(s)} \quad (8.14)$$

By comparing equations (8.4) with (8.14) it is worth noting that the DR tuning for the 2DOF and the 1DOF controllers are the same, i.e. ρ_C will be equal to ρ_I for DR tuning. However the 2DOF controller has an added transfer function which allows the SP to also be optimized, where a 1DOF controller the parameters, ρ_I must be separately tuned for SP and hence it can't cope with both DR and SP simultaneously. This means that there are three controllers to be compared, the 2DOF controller, the 1DOF-DR (tuned to optimize DR) and the 1DOF-SP (tuned to optimize SP).

The 1DOF controller is tuned using IFT as well so the performance of the two 1DOF controllers can be compared with the 2DOF controller independently from the tuning algorithm. As the 1DOF controller is tuned using IFT it also has the ability to adapt to the system parameter drift and converge to an optimal state, but it cannot simultaneously possess the same DR and SP response as the 2DOF controller.

The IFT procedure for the 1DOF controller is the same as laid out in Chapter 5. For the 1DOF-DR controller the reference for the normal experiment is zero and the reference for the gradient experiment is set to $-y^I(\rho)$ so the term in the square brackets in equation (8.15) can be found from this gradient experiment. The terms $\dfrac{1}{G(\rho)}$ and $\dfrac{\partial G(\rho)}{\partial \rho}$ can be calculated and hence can be found. For tuning the 1DOF-SP controller the reference for the normal experiment is a desired set point and the reference for the gradient experiment is the error from the first experiment, i.e. $r - y^I(\rho)$, so now $\dfrac{\partial \widehat{y}(\rho)}{\partial \rho}$ can be found from equation (8.16).

$$\frac{\partial \widehat{y}}{\partial \rho}(\rho_i) = \frac{1}{G}(\rho)\frac{\partial G}{\partial \rho}(\rho)\left[\frac{G(\rho)G_{manip}}{1+G(\rho)G_{manip}H_s}(-y(\rho))\right] \quad (8.15)$$

$$\frac{\partial \widehat{y}}{\partial \rho}(\rho_i) = \frac{1}{G}(\rho)\frac{\partial G}{\partial \rho}(\rho)\left[\frac{G(\rho)G_{manip}}{1+G(\rho)G_{manip}H_s}(r-y(\rho))\right] \quad (8.16)$$

Through this procedure both $\widetilde{y}_i(\rho)$ and an unbiased estimate of $\dfrac{\partial \widehat{y}(\rho)}{\partial \rho}$ can be found for both 1DOF controllers and hence an estimate of $\dfrac{\partial J(\rho)}{\partial \rho}$ can be found, as required for updating the controller parameters.

8.4. Micromanipulator Simulation and Validation

The manipulator system has been simulated in open-loop to validate that it can achieve all the required mechanical specifications, i.e. has required dexterity an ROM etc. and now that the new IFT algorithm has been developed for the precise and robust 2DOF control system the entire device is simulated to validate the control system will work on the manipulator. The full model of the system was developed by integrating the IPMC actuator model and the mechanical model of the mechanism and this was used to simulate the system response and validate the entire design with control system.

8.4.1. Procedure

First the disturbance rejection problem for the 2DOF system was solved through IFT by applying a simulated mechanical disturbance onto the system to mimic a real situation when the manipulator will need to apply a force onto an external unknown environment to carry out an instructed task. The disturbance is modeled by a set impedance which is injected, pushing against the tip of the IPMC with a displacement of 1 mm at $t = 5$ s and then removed at $t = 35$ s. This impedance was chosen to represent a sponge as this can be practically implemented on the IPMC in the real experiments in order to validate the simulation. The real force exerted by the sponge was measured to be proportional to the square of the deflection, this is because as sponge compresses it becomes denser. The peak force at 1mm deflection is approximately 1.5 gf of 15 mN. This is above the typical force needed when dealing with cell manipulations [193, 195]. The reference input is set to zero so the IPMC tries to push back against the spring and then when the spring is removed it overshoots back.

Once the ρ_C and parameters are found the values for ρ_f are tuned by fixing ρ_C and inputting a set point reference of $1°$ at $t = 0$ s then back to zero at $t = 15$ s, then to $-1°$ at $t = 30$ s and then finally back to zero at $t = 45$ s for the last 15 s of the experiment.

The step sizes, γ, for ρ_C were chosen as 0.05 for K_p and K_i and 0.005 for K_d. The step sizes for ρ_f were 0.05 for both α and β. The IFT starts tuning from initial values of $\rho_C = [3000\ 1000\ 100]^T$ which were chosen based on a stable controller that was found using the IPMC model developed in Chapter 3 and some initial experiments. The initial 2DOF parameters were set to, $\rho_f = [0\ 0]^T$ so the tuning starts initially from the DR tuned system.

8.4.2. Simulated DR Tuning

The results are shown below in Fig. 8.11 for the initial un-tuned controller and then for the system after 10 iterations. The shaded area in Fig. 8.11 represents the time period when the disturbance is applied and its magnitude. It can be seen that the IFT tunes the system well.

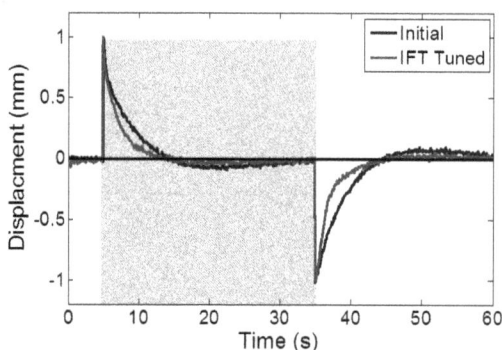

Fig. 8.11. Simulated disturbance rejection response for the initial and IFT tuned systems. The shaded area represents the disturbance input.

Fig. 8.12 (a) shows the cost function, J, for the DR tuning and it can be seen that the IFT algorithm does improve the DR by 37% over the 10 iterations. Fig. 8.12 (b) shows the adjustment of the controller gains over the same 10 iterations.

8.4.3. Simulated SP Tuning

The ρ_C controller gains are fixed and the IFT algorithm is then setup to tune the manipulator system with integrated IPMCs for SP. The results for the tuning the horizontal axis, θ_l, are shown below in Fig. 8.13 (a) and for the vertical axis in Fig. 8.13 (b). It can be seen that the IFT does improve the system response for the SP for both axis.

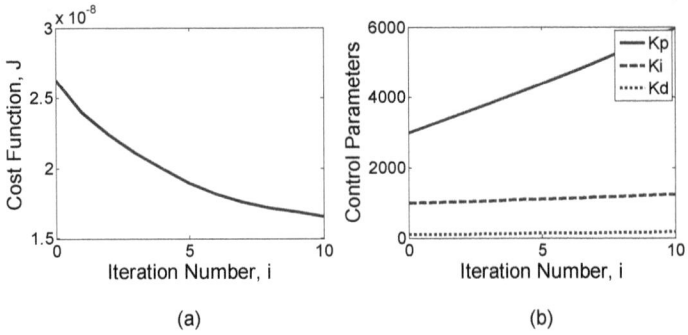

Fig. 8.12. (a) Cost function and (b) controller gain updates for 10 iterations of simulated DR tuning.

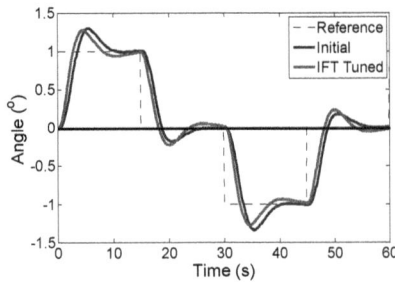

(a)

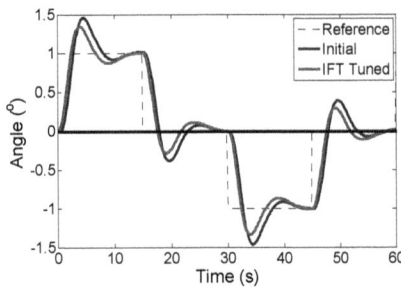

(b)

Fig. 8.13. SP response for the initial and IFT tuned systems for (a) horizontal and (b) vertical direction.

Fig. 8.14 (a) and (b) show the cost functions for the horizontal and vertical motions respectively through the tuning process. It is clear that the IFT is improving the performance in both directions and after 10 iterations the system has improved by 24 % in the horizontal and 28 % in the vertical direction. Fig. 8.14 (c) and (d) shows the adjustment of the 2DOF controller gains, α and β, for the horizontal and vertical motions, respectively over the 10 iterations.

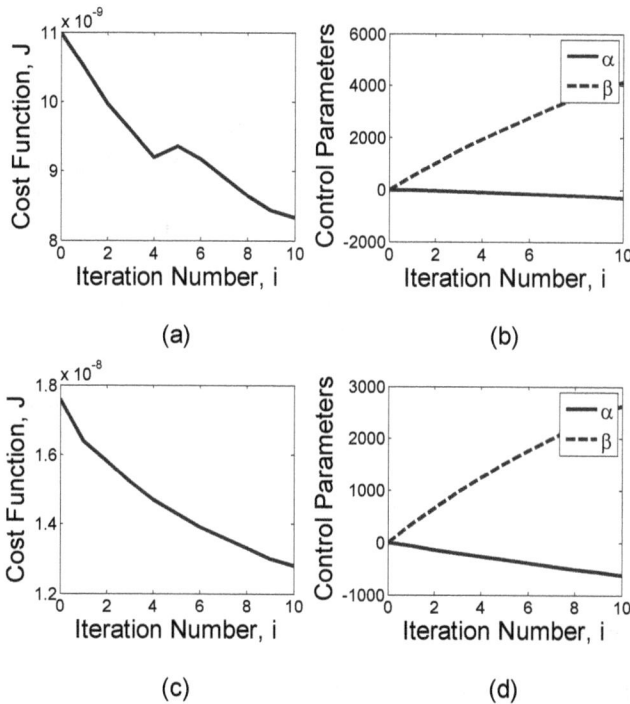

(a)   (b)

(c)   (d)

Fig. 8.14. (a) Cost function and (b) controller gain updates for the horizontal axis, (c) cost function and (d) controller gain updates for the vertical axis for 10 iterations of SP tuning.

From these simulations it is concluded that the robotic manipulator can closely track a desired trajectory as well reject the influence of large external disturbances. The simulations of the IFT algorithm demonstrate that the tuning method does indeed improve the system performance, even with the nonlinear system model. From the results it is clear that the IFT is tuning the system towards an optimal state when using the time-invariant IPMC and mechanism model. It can be

177

assumed that as the tuning is adaptive and based on the actual system performance, when the real system is tuned the controller will adapt its performance even when the IPMC parameters drift. The robotic system and IFT algorithm have been validated that it will meet the specifications 7-9 and so the manipulator is fabricated and experimental work undertaken. First experimental results on the IPMC itself were conducted to test the system for a microtool/gripper, before implementing it on the manipulation system.

8.5.   Microtool/gripper Experiments and Results

To test the robust and precise 2DOF control system in experiments a single IPMC was used first. The IPMC beam represents an IPMC based microtool, which can be used to carry out a number of micromanipulation tasks and also can be attached to the end of the micromanipulator, as previously described in sections 8.2.1 and 8.2.3 respectively. The DR tuning will make sure that the microtool carries out its tasks even in the presence of disturbances. These disturbances could be due to friction or repulsion forces from the cell being moved, reaction force from a cell being gripped or cell being injected, environmental effects or even noise. The SP tuning then ensures that the microtool can follow a desired trajectory very accurately. The combined 2DOF controller will ensure the microtool is both robust and precise.

8.5.1.   IPMC Actuator and Test Setup

An IPMC from ERI based on an XR resin, with Pt electrodes was used for all experiments. The IPMC has a length of 27.5 mm, is 10 mm wide with a thickness of 800 μm. These new IPMC actuators, based on the XR resin, are more immune to the effects of dehydration than the Nafion® based IPMCs, and can in fact operate in air for relatively long periods of time and as such the manipulator system can operate equally as well in air as in aqueous environments. This dimension of IPMC gives a compliance of 0.0277 m/N.

A custom test rig shown schematically in Fig. 8.15 was designed to house the IPMC in cantilever configuration and allowed voltage to be applied at the 5 mm clamped section. A laser sensor with a 10 μm resolution was used to measure the displacement of the IPMC from one side. When the IPMC system is to be used in real micromanipulation operations image processing from a microscope can easily be implemented to replace the need for the external laser sensor for position feedback. A linear motor was used on the other side, with a piece of sponge attached to the end of the shaft to exert a mechanical disturbance to the IPMC. The system was programmed to push the sponge against the IPMC, at specific times causing it to deflect. This was used to simulate a real experiment disturbance. The force exerted by the real sponge is proportional to the square of the deflection, this is because as sponge compresses it becomes denser. Control electronics and a National Instruments DAQ card were used to interface between the MatLab environment running the Simulink model on the PC and the IPMC actuator. The sampling frequency was 100 Hz and the controller was running at 10 Hz.

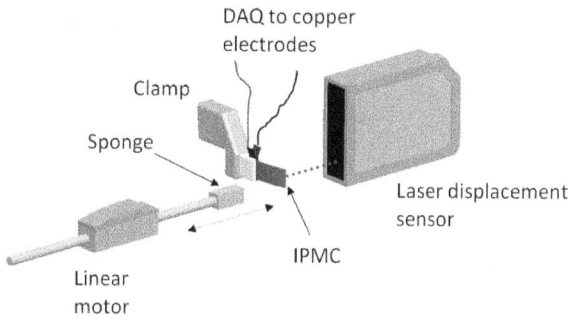

Fig. 8.15. Schematic of experimental setup [220].

8.5.2. Procedure

The same procedure was undertaken for the experimental testing as was done for the simulations as in section 8.4.1. The DR for the 2DOF and 1DOF systems was solved through IFT by applying the real mechanical disturbance onto the system by programming the motor to advance the sponge to push against the IPMC at the specified times causing it to deflect. Once the ρ_C and ρ_I parameters are found the values for ρ_f are tuned by fixing ρ_C and inputting a set point reference of 1 mm at $t = 0$ s then back to zero at $t = 15$ s, then to -1 mm at $t = 30$ s and then finally back to zero at $t = 45$ s for the last 15 s of the experiment. Finally the 1DOF controller is also tuned for SP to give a reference to measure the 2DOF controller against. The results and some verification are presented next.

8.5.3. Results

Fig 8.16 shows the DR for the initial 2DOF and 1DOF-DR controllers and then the final tuned response after 10 iterations of the controller parameters. The response for the two controllers is the same in the DR criteria as their transfer functions, equation (8.3) with $Gf = 0$ and equation (8.13), are the same. The shaded area shows the time and magnitude of the disturbance applied by the sponge. It is clear the IFT has indeed improved the system DR significantly, both when the disturbance is first applied and then the overshoot after the disturbance has been removed. Fig 8.16 (a) shows the improvement of the design criteria over the 10 iterations, the system has improved by 85 %. Fig. 8.17 (b) shows the updates for ρ_C and ρ_I over the tuning experiments.

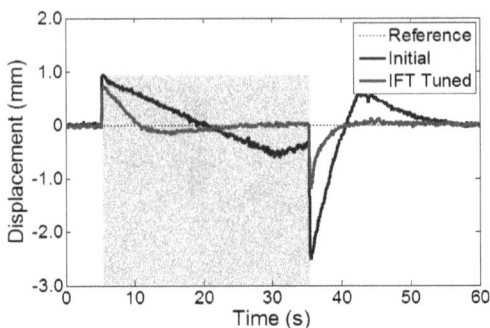

Fig. 8.16. DR tuning results with the initial un-tuned and final IFT tuned 2DOF and 1DOF-DR controller parameters ρ_C and ρ_I respectively.

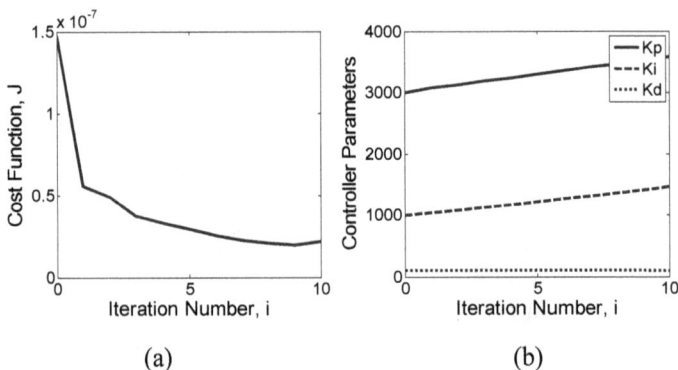

(a)   (b)

Fig. 8.17. (a) Cost function and (b) parameter adaptation for DR IFT tuning of the 2DOF and 1DOF-DR controller parameters ρ_C and ρ_I respectively.

Fig. 8.18 shows the SP tuning for the 2DOF controller which tunes ρ_f with the parameters of G_C fixed at $\rho_C = [3648 \ 1525 \ 99]^T$ as were found previously. It can be seen that the SP performance is tuned to perform much better after the 10 iterations. The improvement in the design criteria is clear from Fig. 8.19 (a), which drastically minimizes after only 1 iteration and then settles at the optimal state. The IFT has improved the SP of the 2DOF controller by 69% from the DR tuned system. The updated controller parameters ρ_f are shown in Fig. 8.19 (b).

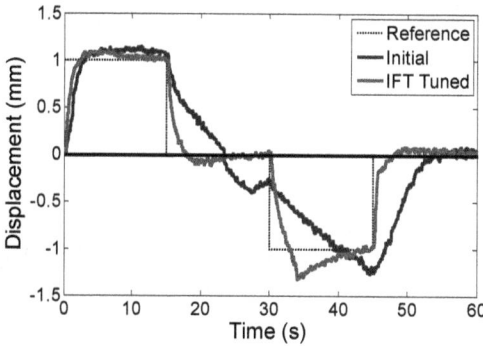

Fig. 8.18. SP tuning results with the initial and final 2DOF control parameters ρ_f.

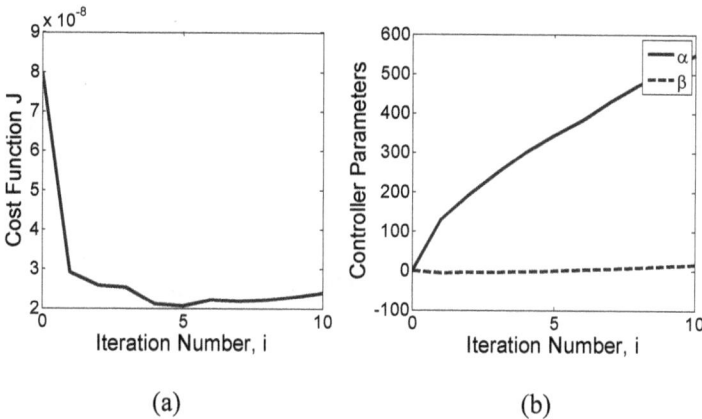

(a)   (b)

Fig. 8.19. (a) Cost function and (b) parameter adaptation for SP IFT of the 2DOF control parameters ρ_f.

Finally the SP is tuned for the 1DOF-SP controller, the results which show the initial un-tuned and the final tuned response after 10 iterations are presented in Fig. 8.20. The IFT algorithm successfully manages to improve the SP response of the system by 70 % as can be seen in Fig. 8.21 (a). Fig. 8.21 (b) shows the updates of ρ_l as the system is tuned towards an optimal state.

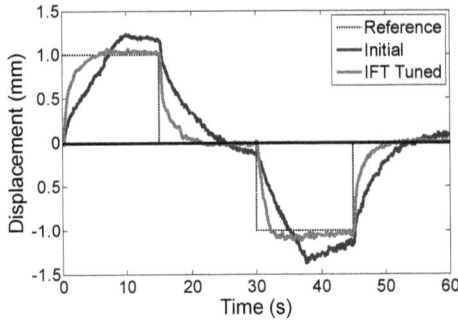

Fig. 8.20. SP IFT tuning results with the initial and final 1DOF control parameters ρ_l

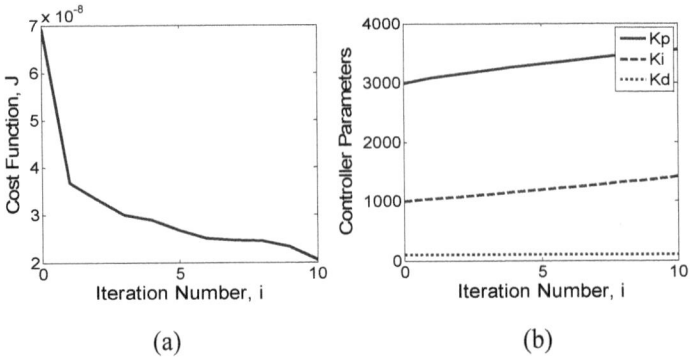

(a)   (b)

Fig. 8.21. (a) Cost function and (b) parameter adaptation for IFT of the 1DOF-SP control parameters ρ_l.

8.5.4. Validation of 2DOF Controller

The 1DOF-SP controller is tested for the DR and then compared to the 2DOF and 1DOF-DR controllers to validate that the controllers have been tuned appropriately, the results are shown below in Fig. 8.22. It can be seen that the 2DOF and 1DOF-DR controllers perform much better than the 1DOF-SP controller, as expected. When the disturbance is applied at $t = 5$ s, the DR tuned systems perform better by instantly opposing the force applied and then they continue to outperform the 1DOF-SP system while the disturbance is applied, $5 < t < 35$ s, by settling to the reference more quickly. Once disturbance is removed at $t > 35$ s, the SP tuned and DR tuned controllers show comparable results. This is expected as once the disturbance has been removed the system will now simply following the reference trajectory in free deflection, as there is no longer any disturbance.

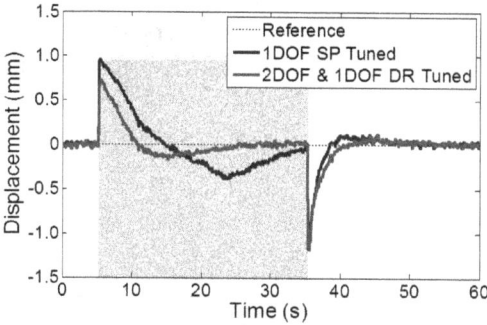

Fig. 8.22. Comparison of the 2DOF and 1DOF-DR with the 1DOF-SP for the 1 mm disturbance input.

Next the systems are validated for SP tuning by comparing the 2DOF controller with the 1DOF-SP and 1DOF-DR controllers as shown in Fig. 8.23. It can be seen that the tuned 2DOF and 1DOF-SP are fairly comparable to each other. The 2DOF controller does tend to overshoot more than the 1DOF-SP controller but the settling times are the same for both. Both controllers far outperform the 1DOF-DR controller for SP tracking.

As further validation the 2DOF controller and the two 1DOF controllers were tested in a mix of tracking with a disturbance as well. The same SP reference as the previous experiment is input while the sponge was fixed at the zero displacement point. This means that on the positive side, as shown in Fig. 8.24 there is free deflection while on the negative side there is a disturbance when the system is attempting to track the reference. The results are shown below and the shaded area represents the side of the disturbance. It can be seen that initially for the free deflection, $t < 50$ s, the 2DOF and 1DOF-SP perform fairly

183

comparably, (2DOF overshoots slightly more, but settles more quickly) while the 1DOF-DR is slower to rise, takes longer to settle and overshoots. When $t > 50$ s, it can be seen that the IPMC takes a lot longer to reach the SP for all of the controllers, as expected because it has to push against the sponge. Also there is a lot of high frequency oscillation of the IPMC which is caused by the controller trying to push the sponge out of the way. It is observed that although the 2DOF controller acts slightly more slowly than the 1DOF controllers to reach the negative SP, it actually exhibits a lot less oscillation in this period than both of the 1DOF controllers. When the reference is set back to 0 at $t = 75$ s, the 2DOF controller again acts more slowly, but seemingly is steadier than the 1DOF controllers when approaching the zero displacement mark. The 1DOF controllers both overshoot the zero displacement by a lot more than the 2DOF controller and they all seem to settle around the same time.

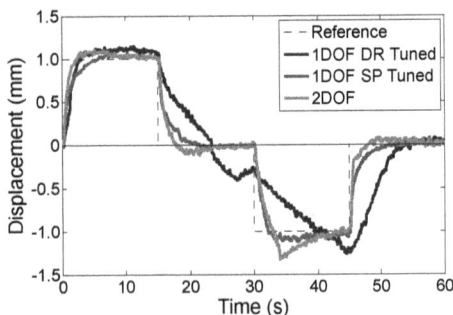

Fig. 8.23. Comparison between the 2DOF controller and the 1DOF-DR and 1DOF-SP tuned controllers for the SP response.

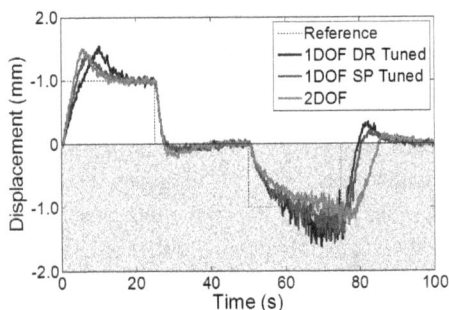

Fig. 8.24. Comparison between the 2DOF controller and the 1DOF-DR and 1DOF-SP controllers for a mixed set point tracking and mechanical disturbance.

8.6. Micromanipulator Experimental Results

The final design for the robotic manipulator, which has been validated through simulations, was rapid prototyped using ABS material as shown in Fig. 8.25. Steel pins which are lubricated are used for the two rotary joints to reduce the friction in the system. The IPMCs are clamped into place with cooper electrodes, which are plugged into the connection block (where the colored cables are wired to) with extremely thin and lightweight electrical cable to minimize the disturbances to the manipulator during operation. The connector block is wired to the data acquisition card which sends inputs and outputs to the control system which is implemented in Simulink®.

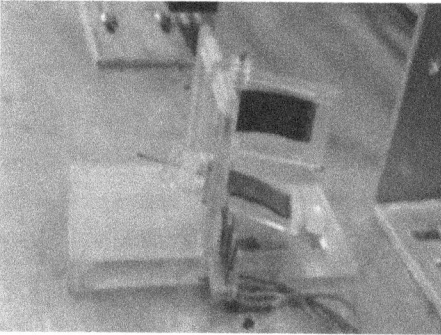

Fig. 8.25. Manipulation system module with IPMCs and micro-probe [220, 221].

Experiments are undertaken on the system to tune the manipulator in both joint axes, for both the 2DOF and 1DOF controllers. Once the controllers have been developed experiments to measure the range and accuracy are undertaken and finally some tracking references are input which combine both horizontal and vertical motions to validate the systems performance. A demonstration of the manipulator response when pushing and lifting a small aluminum block is also presented to show the superior performance of the 2DOF controller in comparison with 1DOF controller.

8.6.1. Procedure

The two step tuning procedure as described previously is undertaken using IFT to first optimize the DR of the IPMC and then the SP tracking of the manipulator. In order to tune for DR a repeatable mechanical disturbance must be injected at the same time in each experiment. The most accurate way to carry this out practically is to inject the disturbance on to the IPMC separately, without the mechanism. This then implies that the IPMC actuator itself is tuned for DR and when integrated into the manipulator the load applied by the mechanism arms is a disturbance on to the IPMC. This can be justified as the manipulator dynamics add a relatively small load onto the IPMC, in both the horizontal and vertical direction. The disturbance loads will be due to inertia, Coriolis effect, gravity and friction of the manipulator. Gravity compensation has been implemented, also the mass, velocities and accelerations are extremely low causing the inertial and Coriolis torque to be low as well as the friction, which is dependent on the normal reaction force.

The setup for the DR experiment is the same as for the DR tuning for the microtool/gripper. A laser sensor was used to measure the displacement of the IPMC from one side and a linear motor was used on the other side, with a piece of sponge attached to the end of the shaft to exert a mechanical disturbance to the IPMC. This is repeated for the two IPMC strips for the horizontal and vertical axes. The parameters need to be tuned for each IPMC as the materials are different for each different strip, due to the IPMC manufacturing process. Once the ρ_C parameters are found for the two IPMCs after DR tuning, the IPMCs are mounted into the manipulator, as per Fig. 8.25 and the values for ρ_f are then tuned by fixing ρ_C and setting a combination of $\pm 1°$ set point references. Finally the 1DOF controller is also tuned to give a reference to compare the 2DOF controller against.

The step sizes, γ, for ρ_C and also the 1DOF controller were chosen as 0.05 for K_p, K_i and 0.005 for K_d. The step sizes for ρ_f were 0.05 for both α and β. The IFT starts tuning from initial values of $\rho_C = [3000 \ 1000 \ 100]^T$ which were chosen based on a stable controller found using the IPMC model for the IPMCs and some initial experiments. The initial 2DOF parameters were set to, $\rho_f = [0 \ 0]^T$ so the tuning starts initially from the DR tuned system.

8.6.2. DR Tuning

The disturbance response for the IPMC, which will be used for the horizontal motion in the manipulator, is presented in Fig. 8.26. The shaded area in Fig. 8.26 represents the time period when the disturbance is applied and its magnitude. It can be seen that the IFT tunes the system well as the displacement for the initial un-tuned controller shows much larger errors than the IFT tuned controller. The tuned controller reacts more quickly when the disturbance is first applied, restricting the motion, and then also reacts more quickly when the disturbance is removed, returning quickly to the zero reference.

Fig. 8.27 (a) shows the cost function, J from equation (5.27), for the DR tuning and it can be seen that the IFT algorithm does improve the DR by 85 % over the 10 iterations. Fig. 8.27 (b) shows the adjustment of the controller gains over the same 10 iterations.

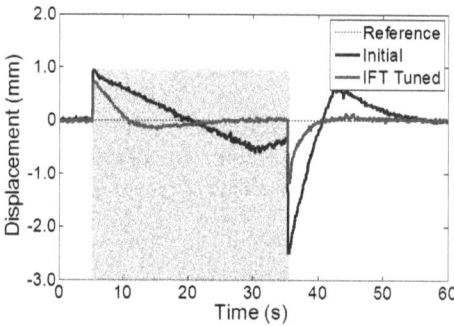

Fig. 8.26. Disturbance rejection response for the initial and IFT tuned IPMC for the vertical motion axis, θ_1. The shaded area represents the disturbance input.

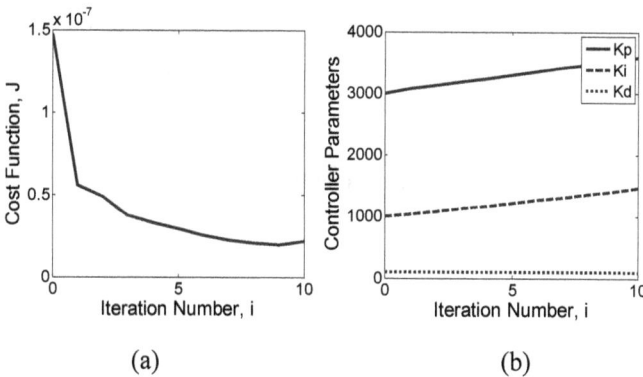

(a)   (b)

Fig. 8.27. (a) Cost function and (b) controller gain updates for 10 iterations of DR tuning for the horizontal motion axis, θ_1.

The same approach was undertaken for the IPMC to be used for the vertical motion in the manipulator, see Fig. 8.28. It was again found that the IFT algorithm tunes the DR well with an 87 % performance increase as seen in Fig. 8.29 (a). Fig. 8.29 (b) shows the ρ_C updates over the 10 iterations.

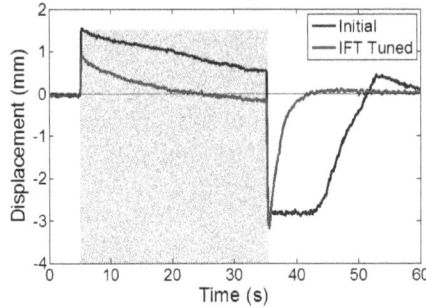

Fig. 8.28. Disturbance rejection response for the initial and IFT tuned IPMC for the vertical motion axis, θ_2. The shaded area represents the disturbance input.

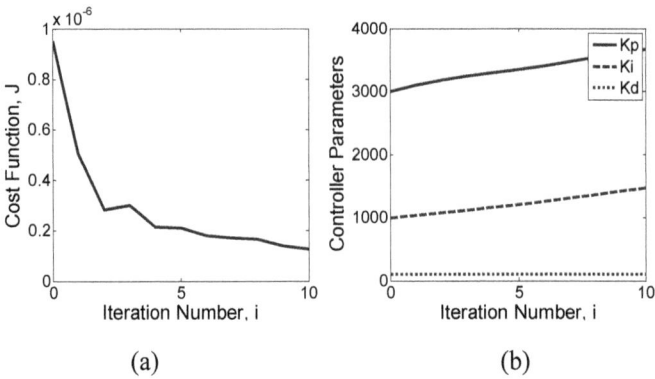

| (a) | (b) |

Fig. 8.29. (a) Cost function and (b) controller gain updates for 10 iterations of DR tuning for the vertical motion axis, θ_2.

8.6.3. SP Tuning

With the results for the controller gains, ρ_C, fixed the IFT algorithm is set to tune the manipulator system with integrated IPMCs for SP. The

results for tuning the horizontal axis, θ_1, are shown in Fig. 8.30 (a) and
for the vertical axis in Fig. 8.30 (b). It can be seen that the IFT does
improve the system response for the SP for both axes.

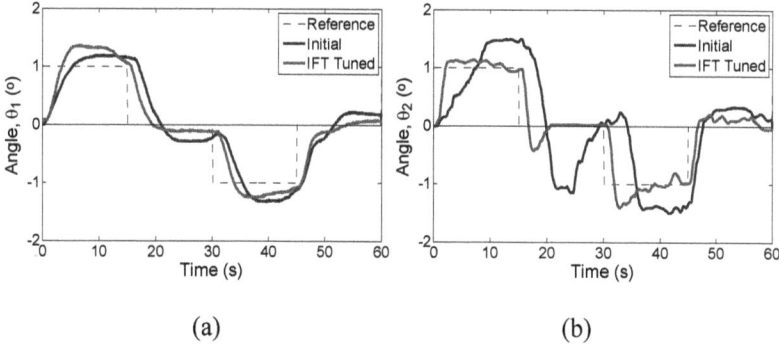

(a) (b)

Fig. 8.30. SP response for the initial and IFT tuned systems
for (a) horizontal and (b) vertical direction.

Fig. 8.31 (a) and Fig. 8.31(c) show the cost functions for the horizontal
and vertical motions respectively through the tuning process. It is clear
that the IFT is improving the performance in both directions and after
10 iterations the system has improved by 33 % in the horizontal and
81 % in the vertical direction. It is clear from these figures that the IFT
quickly improves the performance and tunes itself towards an optimum
in only 10 iterations Fig. 8.31 (b) and Fig. 8.31 (d) show the adjustment
of the 2DOF controller gains, α and β, for the horizontal and vertical
motions, respectively over the 10 iterations.

The results from the SP tuning experiment with the manipulator in
Fig. 8.30 and Fig. 8.31 show that the horizontal direction performs
better than the vertical. It is very noticeable that the vertical direction
has a lot more 'jerky' response than in the horizontal direction which is
very smooth. This can be explained as the weight of the bottom arm
acts in the normal direction to the motion of the pin-arm rotating
surface of the rotary pin joint #2, this directly affects the friction
(Coulomb friction model, $F=\mu.N$) as seen in Fig. 8.32. This explains
why the motion takes longer to get started and then once the static
friction is overcome, the build-up of the IPMC force and arm inertia
causes the position to overshoot a lot (stick-slip phenomena). Also the
IPMC used in the vertical motion has to overcome some gravitational
effects. In the horizontal axis, the weight of the arm acts parallel to the

pin, so the normal force which causes the friction in the joint is minimal, hence the smooth operation in that axis.

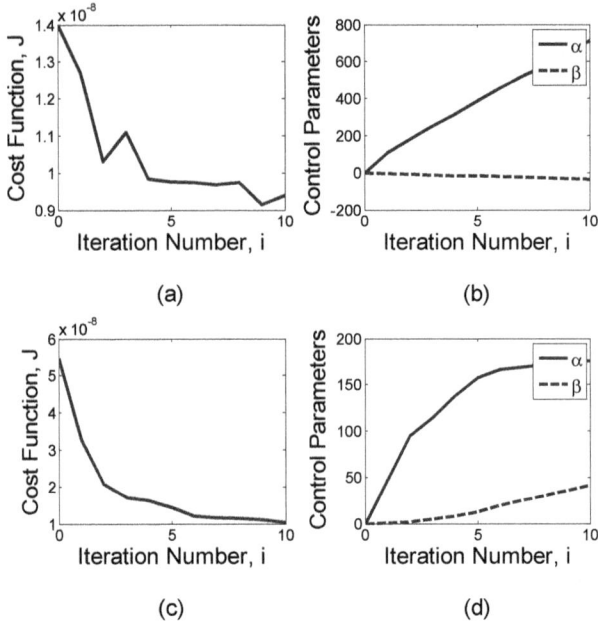

Fig. 8.31. (a) Cost function and (b) controller gain updates for the horizontal motion. (c) Cost function and (d) controller gain updates for vertical motion for 10 iterations of SP tuning.

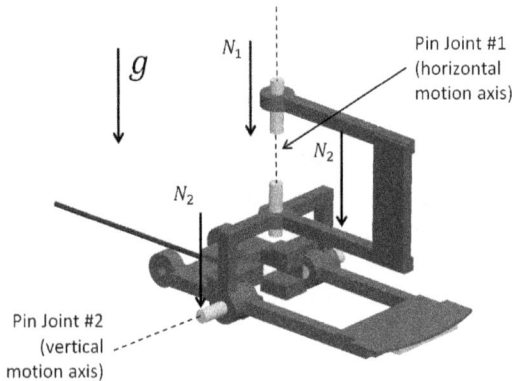

Fig. 8.32. Schematic drawing of pin joints and normal forces which cause friction.

Despite this it can be seen that the IFT does improve the system response very well in the vertical direction and actually reduces the overshoot significantly over the 10 iterations. In fact the tuning procedure works better in the vertical direction with a larger performance improvement.

8.6.4. 1DOF Tuning

Output results for tracking the SP reference in the horizontal axis, θ_1, for the initial unturned controller and the final IFT tuned controller after 10 iterations are shown in Fig. 8.33 (a). The cost function over the tuning procedure is plotted in Fig. 8.33 (b).

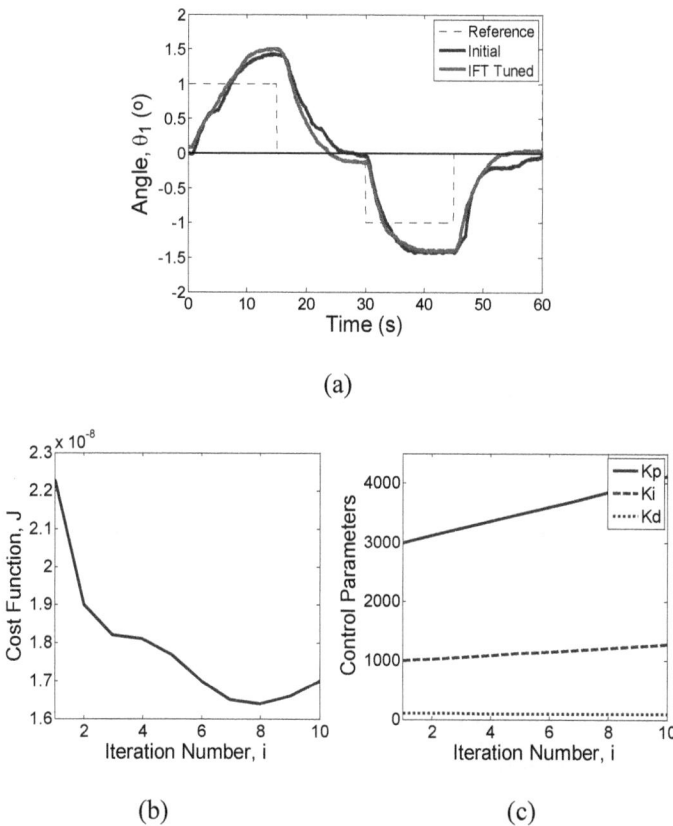

(a)

(b)

(c)

Fig. 8.33. (a) Horizontal tracking response for a set point reference for the initial and IFT tuned systems. (b) The cost function and (c) controller gain updates for 10 iterations of IFT tuning.

From this it is clear that the IFT does improve the system response over the 10 iterations, the performance is improved by 24 %. It should be noted however that for iterations 9 and 10 the cost is actually increasing, this may be due to the controller overstepping the optimum parameters, if this is the case the system would adjust in future iterations to come back to this optimum. It is more likely though that this can be explained due to the highly time-varying nature of the IPMC samples. Over a large number of iterations it is expected that there will be some random outliers as is seen here. Despite this, overall the system response has been improved. Fig. 8.33 (c) shows the parameter update over the tuning procedure.

The results for set point tracking for the initial and IFT tuned controllers for the vertical direction are shown in Fig. 8.34(a). The first thing which is noticeable is that the vertical direction is much slower and overshoots more than the horizontal motion. This more jerky response can be explained through the same logic as in the previous section for the 2DOF controller. Despite this it can be seen that the IFT does improve the system response very well and actually reduces the overshoot significantly in 10 iterations. Fig. 8.34 (b) shows the cost function over the tuning experiments and it can be seen that the performance is improved significantly after even 2 iterations. Overall the system improves by 64%. The parameter updates are shown in Fig. 8.34 (c).

It can be seen from the tuning results that it is extremely difficult to control both arms to follow a reference trajectory perfectly due to the complex behaviors of the IPMCs themselves as well as the interaction between the IPMCs and the mechanical mechanism. Although the system performance in both axes is not perfect with the simple 1DOF PID controller, the IFT has successfully worked to improve the systems performance. Also it can be noted that a longer time period than 15 s should be given to the system so it is guaranteed to settle at the requested target value.

8.6.5. Comparison of IFT Tuned 2DOF and 1DOF Controllers for Tracking

With both the 2DOF and 1DOF controller tuned for each axes of the manipulator their performance is compared, in terms of tracking response. As both systems have been tuned for 10 iterations and are close to their optimal state, the two controllers can be compared fairly independently of the tuning algorithm, hence only the controller structures are being compared. The final tuned 1DOF and 2DOF controllers are compared in Fig. 8.35 where it can be seen that the 2DOF controller does outperform the standard 1DOF controller as expected. The 2DOF controller is 49 % better in horizontal and 76 % better in the vertical axis. Again it is seen that for both controllers that the horizontal motion is much smoother and this shows that it is a mechanical issue, rather than a problem with the controllers themselves.

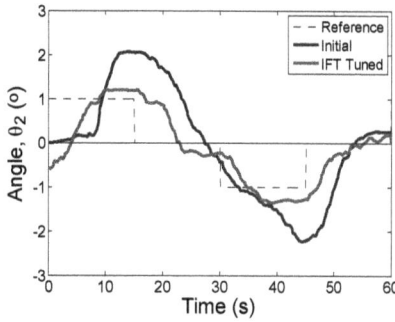

(a)

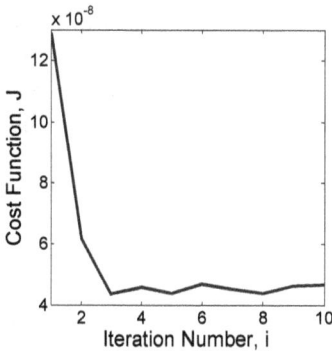

(b)

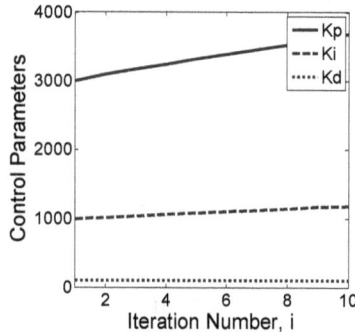

(c)

Fig. 8.34. (a) Vertical tracking response for a set point reference for the initial and IFT tuned systems. (b) The cost function and (c) controller gain updates for 10 iterations of IFT tuning.

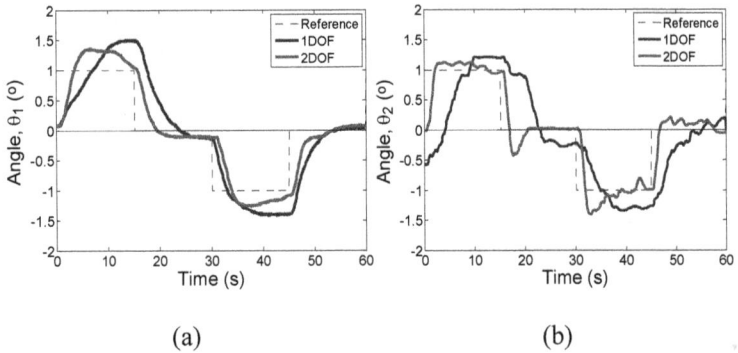

(a)   (b)

Fig. 8.35. SP response for the 1DOF and 2DOF controllers after 10 iterations of IFT in the (a) horizontal and (b) vertical direction.

8.6.6.   Range and Accuracy

Fig. 8.36 (a) and (b) show experiments which were conducted to determine the range and accuracy for the manipulator in the horizontal and vertical axes. The reference was a step of $0.5°$ every 25 s. From the results it is clear that the manipulator does track the reference in both DOF. Again it is noticeable that the horizontal axis performs much better and more smoothly than the vertical axis as is seen in the tuning experiments. The vertical is much more 'jerky' and overshoots more than the horizontal motion. This phenomenon is seen for both the 1DOF and 2DOF controllers and so this eliminates the possibility it is the controller causing this behavior and thus validates the hypothesis that it is caused by the mechanism itself, mainly caused by the higher friction as well as the gravitational effects as explained in the previous section.

In the horizontal direction the 1DOF and 2DOF controllers are almost identical. This is explained as there are minimal disturbances in this motion axis and as such the DR characteristics of the 2DOF are not fully exploited. This shows that the mechanism adds little load onto the IPMC in the horizontal axis.

In the vertical direction the performance for both controllers is comparable up to $3.5°$ and after that the 1DOF controller performance degrades significantly compared to the 2DOF controller which accurately tracks all the way to $7°$. This is expected as at higher displacements the available IPMC force decreases so the effects of friction become more prominent as the IPMC struggles to overcome them. At high displacements, as the friction disturbance becomes large compared with the available force output it is seen that the 2DOF controller, with good DR, outperforms the 1DOF controller.

The 2DOF controllers can accurately achieve up to 7° in each joint axes. Both the horizontal and vertical motion reference is tracked to within < 0.1° error which is 1.4 % of full scale.

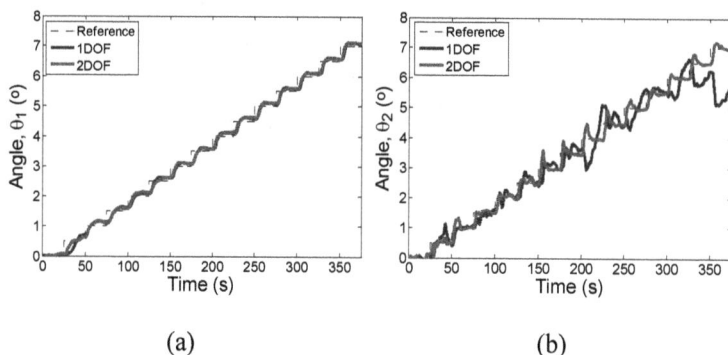

(a)   (b)

Fig. 8.36. Stair step input to measure the range and resolution for the (a) horizontal and (b) vertical direction [220].

8.6.7. 2 - Axis Tracking Performance

To measure the performance of the manipulator with combined axis motion three shapes were input as reference signals, a 2° diameter circle, a square with 2° sides and a diagonal reference up to the maximum displacement of 7° in both axes simultaneously. From the results in Fig. 8.37 it is seen that both IPMCs can operate well together to achieve a desired tracking response.

The major errors for both the 1DOF and 2DOF controllers come from the vertical direction, as has been seen in the previous experiments. The vertical motion produces some overshoot and oscillation, while the horizontal motion is well behaved.

From the results it is clear that the 2DOF controller outperforms the 1DOF controller in both the circle and square plot trajectories. This validates that the 2DOF controller performs better than the 1DOF controller for position tracking in 2 joint axes.

The maximum error of the 2DOF controller in combined 2 axis motion is 0.5° which corresponds to 7.1 % of full scale.

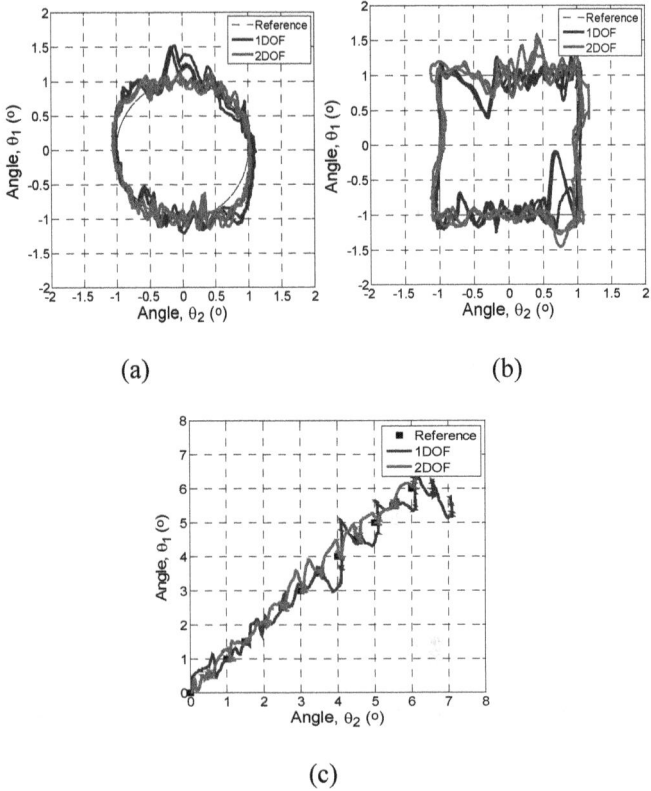

(a)   (b)

(c)

Fig. 8.37. Combined 2 axes tracking for a (a) circle, (b) square and (c) diagonal line [220].

8.6.8. Tracking Performance with an External Disturbance

It has been shown that the 2DOF controller outperforms the 1DOF controller when the manipulator is in free motion and the IPMC only has to move the manipulator arms. In real operation there will be other external loads, for instance when the manipulator probes a cell or has to move a micro object. The robustness of the manipulator to these types of external disturbances must therefore be validated. This was done by making the manipulator first push a 1 g aluminum block along the worktable in the horizontal direction and then lift this aluminum block in the vertical direction. The results for the 1DOF and 2DOF controllers are presented in Fig. 8.38.

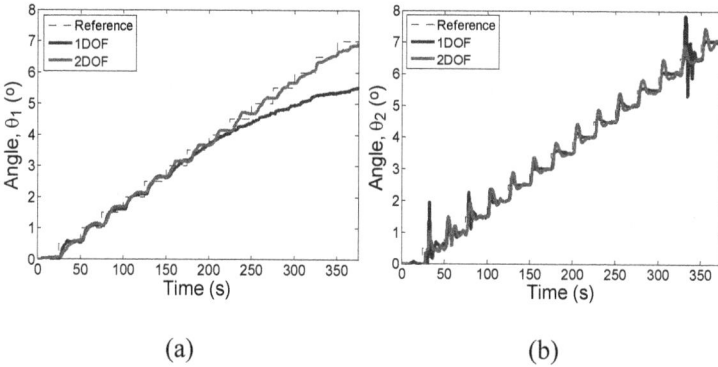

(a)   (b)

Fig. 8.38. (a) Pushing and (b) lifting the 1 g block [220].

From these experiments it is clear that the 2DOF controller outperformed the 1DOF controller. In the horizontal direction the manipulator cannot keep up with pushing the block due to the friction force between the block and worktable. After approximately 4° the 1DOF lags well behind and stops tracking the reference. This is due to the fact that as the IPMC bends further the available force decreases and as such the disturbance becomes a much larger factor compared to the case of free motion in which the 1DOF controller was tuned.

At the beginning of the vertical lift there is a large overshoot in two of the target references with the 1DOF controller. This is caused again by the stick-slip phenomena, but with the added inertia from the block being lifted, this effect is magnified, i.e. once the arm starts to move it does so quickly and then it is even harder to stop, causing the large overshoot. The same phenomena is seen at high displacement target of 6.5°, where again the IPMC begins to move and with the increased inertia the 1DOF controller finds it difficult to stop the oscillatory motion. The 2DOF controller with its DR characteristics is more capable of handling this added inertia. The stick-slip becomes more important at large displacements as the available force output of the IPMCs decreases with deflection.

In both motions it is seen that the 2DOF controller does cope with the large disturbance very well and the performances are even comparable with the free motion in Fig. 8.36 which suggests that the disturbances are fully accounted for. This has proven that the 2DOF controller is indeed precise and robust up to 7° of motion.

8.7.   Discussion

A novel robotic manipulator system has been developed to successfully meet all the required specifications for cell manipulation, as laid out in section 8.2.2.1. This was achieved by carrying out a thorough design process, from first developing a design based model for the IPMC actuators, then designing a novel robotic mechanism, carrying out simulations in order to tweak the design to get the optimum performance for the available IPMC material before finally building and verifying the design in real life. This manipulator design is, to the best of the author's knowledge, the first example of utilizing standard IPMCs to actuate an external mechanical robotic mechanism to achieve an output in 2DOF. This research also demonstrates that carrying out a thorough design process when developing systems with integrated IPMCs is a very useful task in order to optimize the systems performance.

A simulation analysis was undertaken with all possible configurations of IPMCs that could be cut from the sheet of IPMC material which was available to achieve an optimal system. If different amounts or types of IPMC materials were available the same process can be undertaken to decide the configuration to give a desired performance. This has demonstrated the type of design analysis which can be undertaken when using scalable mechanical design based IPMC models for inventing novel systems, with integrated IPMC actuators for new applications. This procedure is an example of the design process that was followed, but with the scalable IPMC model any number of configurations and sized IPMCs can be simulated for any application.

The manipulator has been designed to have inherent compliance through the IPMCs themselves; this adds some passive safety to the design, for example if the controller causes the microtool to overshoot the cells may still avoid harm where other rigid systems will certainly damage the cells. Despite this the manipulator can still cause damage as it is only controlled with positional feedback and so the system has no knowledge of the force being applied to the environment. If for example there is a variation in the environment or even some controller errors the microtool may come into contact with a cell unexpectedly causing the position error to wind-up, this will result in a high input voltage to the IPMC causing a critical force onto the cell. This can be avoided by lengthening the microtool to reduce the maximum available force that can be exerted so it is below a level which can damage a cell. In this way if there is any unplanned contact, the manipulator cannot provide enough force to harm the cell. Although the system has been designed to be robust in its operation, it is not sufficient to have only position control for extremely sensitive procedures and hence is very important to have force feedback during experiments to ensure safe handling of the cells. With force information active compliance control and fail-safe measures can be employed to develop a truly safe system which has both active and passive safety measures.

Safe manipulation and handling of cells is an extremely difficult task due to the need for precision control coupled with a sensitive touch to carry out delicate tasks. Because these tasks require small tolerances and high specifications no device can currently meet all the requirements in cell manipulation. The device in this research has been developed to overcome many of the shortcomings of current devices, but it is useless without a precise and robust control system to ensure it is accurate and reliable throughout its operation. In this research the developed control system has demonstrated impressive results validating that it can reliably control the micromanipulator to follow any specified trajectory requested by the operator, with precision even in the presence of high disturbances.

It has been shown that the 2DOF controller performs the same as the 1DOF-DR controller when a disturbance is applied and much better for when following a SP. The 2DOF controller shows comparable performance with the 1DOF-SP controller for following a reference input in free deflection and far outperforms it in DR. From these results it can be implied that the 2DOF will have a better overall performance than either of the 1DOF controllers. This has been demonstrated further by showing that the 2DOF controller outperforms either the 1DOF-SP or 1DOF-DR tuned controllers for an overall mixed response of the system under different operating conditions, Fig. 8.22, Fig. 8.23 and Fig. 8.24. This proves that the 2DOF controller can be implemented for cell manipulation in unknown environments to perform precise tracking of a desired reference while remaining robust by rejecting external disturbances.

When comparing the experimental results for the 1DOF and 2DOF controllers it is very clear that having 2 independent transfer functions to regulate two properties of the system greatly improves the performance. When the IPMC has to move the manipulator with no other loads the controllers perform comparably, this proves that the manipulator arms add little disturbance to the system and means that tuning the DR separately from the manipulator is justified. When the 1 g mass, which is much larger than will typically be encountered in cell manipulation, is added to the system the 1DOF controller cannot cope while the 2DOF controller performs almost as if there is no load and the system is in free motion. Based on this observation alone the system has demonstrated that it is both robust and precise in the presence of large disturbances.

From the experimental tracking results it is clear that the horizontal control of the manipulator with a simple PID controller is very accurate, where on the other hand the vertical motion has some significant errors. The errors in the vertical direction can be put down to the necessity of the bottom IPMC to overcome the friction and associated stick-slip phenomena which is higher than in the horizontal direction mainly due to the weight of the arm acting normal to the pin-arm moving surface. It has been noted in experiments that lubricating this joint does help somewhat. It is also expected that if the manipulator was fabricated more precisely than rapid prototyping it from ABS material, then this would reduce the vertical positional error significantly.

Both of the IPMCs used in the manipulator were tuned separately as the dynamics of both axes are different and even the individual IPMCs cut from the same sheet can exhibit different behavior. As the system has an automatic adaptive controller this tuning is very easy, unlike for example a model based controller where a comprehensive system model would be needed to tune each axis. As the system automatically tunes, the manipulator can be easily tuned to optimize its performance for different tasks, for example injection and pushing, as these will require a very different system performance. In this way a number of optimal controller parameters can be stored and then loaded into the controller for the different operating modes when the user specifies the tasks they are undertaking. Also as the system tunes online, the more the system operates the better it gets, the controller is therefore learning how to operate best for specific tasks. This would be complex and time consuming to achieve with traditional model based control.

As the IFT algorithm automatically tunes it is simple to switch parameters for different operating modes. For example if the manipulator is loaded then switch the parameters to this mode, carry out the task and tune the parameters during operation for that mode. Similarly when unloaded, run the task and tune for unloaded situation. A gain schedule, as in Chapter 6 can also be implemented to improve operation.

In the experiments a microtool with a length 20 mm was used. By using different length microtools different force outputs and displacements can be achieved. Also a number of the manipulators can be used together, cooperating to give the system more dexterity.

8.8.   Micromanipulation Summary

A 2DOF micromanipulation system which can be used for the safe handling of sensitive biological materials, due to the inherent compliance of the IPMC actuators has been developed and implemented. The system was first designed in simulation utilizing a developed IPMC model and mechanical model for the robotic mechanism to refine the performance before implementing the system for real application.

A model-free IFT algorithm has been proposed to adaptively tune both the DR and SP of the system. This is accomplished by introducing a 2DOF control structure which will allow optimal performance under both measures where a traditional 1DOF controller has to either be tuned for the DR or SP and cannot achieve an overall optimum. Also by developing an adaptive tuning system the controller will handle the drift in IPMC dynamics so the system can indeed cope with the time-varying characteristics and hence operate for a prolonged period of time.

Experiments have validated that the manipulator system meets all the desired specifications. This has demonstrated that IPMCs are a unique solution for the delicate tasks required for biological cell micromanipulation. Through this research a number of shortfalls with current cell manipulation devices have been overcome. It is anticipated that this will eventually aid in further advances in the medical and bioscience research fields.

Chapter 9

Force Compliant Surgical Robotic Tool with IPMC Actuator and Integrated Sensing

Robot aided surgery and robotic surgical instrumentation are some of the fastest growing applications in robotics. Many robotic devices have now surpassed human capabilities in a number of ways. One particular example is the daVinci® surgical robotic system from Intuitive Surgical, with more than 2585 systems already installed in hospitals worldwide [212]. Robotic assisted surgery is becoming widely adopted by surgeons for a number of reasons, including advantages such as improved dexterity with advanced instrumentation, undisputed higher precision which leads to shorter learning curves [213], embedded force/torque sensors which have the ability to give direct kinesthetic feedback to surgeon [214], more safety with active constraints through appropriate control algorithms and reductions in surgeon tremor through low pass filtering as well as faster patient recovery times and cosmetic advantages. As a result surgical tasks which were previously considered far too risky for robots are now common practice.

Although current advanced surgical devices like the daVinci® have many advantages they are require hugely complex mechanisms and highly geared servo motors to ensure extremely high accuracy and are therefore mechanically bulky, carry huge price tags and have large end point impedance. The complexity of the instruments makes it difficult to be miniaturized and scaled. IPMCs, on the other hand, are inherently compliant [5] and would therefore have low tip mechanical impedance making the tools inherently back-drivable, which will ensure the safety during the surgery. IPMCs are also easily miniaturized which can be exploited for designing endoscopic or minimally invasive surgical tools. However, much work is still required to overcome some of the downfalls of IPMCs, including hysteresis and non-repeatability, in order to advance them into real world systems. Little work has been done using IPMCs as actuators in rotary joint devices and especially in a surgical tool application. A big downfall of IPMCs has been the need for bulky external sensing (usually load cells or laser sensors) to get accurate feedback, but in this work embedded sensing for accurate force control is achieved. With this accurate force sensing feedback control is implemented, which in addition to the natural force compliant characteristics of the IPMC make a safe and accurate rotary surgical tool.

9.1   Surgical Tool Design

The device will be designed as an end effector to undertake precise and small scale cutting procedures and it will be capable of being attached to a larger robotic arm as shown in Fig. 9.1. The rigid arm will carry out the large (macro) position movement while the IPMC actuated end effector cutting tool will handle the delicate and precise movement (micro) to achieve accurate surgical cutting procedures. This gives the best performance and properties needed for a surgical device.

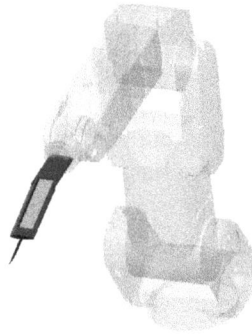

Fig. 9.1. A robotic arm with the proposed end effector as surgical tool [222].

9.1.1.   Mechanical Design

The device is designed as a robotic end effector cutting tool, which will be attached to a larger robotic arm with a bigger working envelope such as the daVinci® system. The robotic arm will bring the IPMC actuated end effector towards the tissue to be cut and the IPMC will handle the precise movements to achieve accurate surgical cutting procedures.

The IPMC mechanism designed is a single degree of freedom (1DOF) device which consists of a 'skeleton' frame with one rotary joint actuated by IPMC, Fig. 9.2. The IPMC is sized to 10 mm × 30 mm × 1 mm with one end clamped and fixed at the joint while the other end can slide in a slot. This actuation method is proposed and proved to be effective in previous work by McDaid *et. al* [215]. The advantage of this mechanism is, in the cutting direction the device will remain complaint while in all the other directions the system is considerably rigid, ensuring all cutting is precise and requires simple sensing and control. As the IPMC is in a rigid frame it is also free from the influence of the irregular deformation of the IPMC itself and load from any undesired direction. In Fig. 9.2 a frame has been used to hold the cutting tool so that it can be used to support the tool for testing, although as mentioned earlier this tool can be connected directly to a larger robot, for example a serial robot as proposed in [215]. This means there is a precise but rigid structure connecting the scalpel actuated by a soft and compliant IPMC actuator.

A soft cantilever beam made from a printed Acrylonitrile butadiene styrene (ABS) plastic connects the scalpel holder and rigid skeleton frame that houses the IPMC actuator. The ABS cantilever is used to give some small, but measurable deflection when a force is applied to the cutting tip. Using an appropriate strain gauge and sensing circuit this bending displacement can be measured and used to determine the cutting force at the tip of the scalpel.

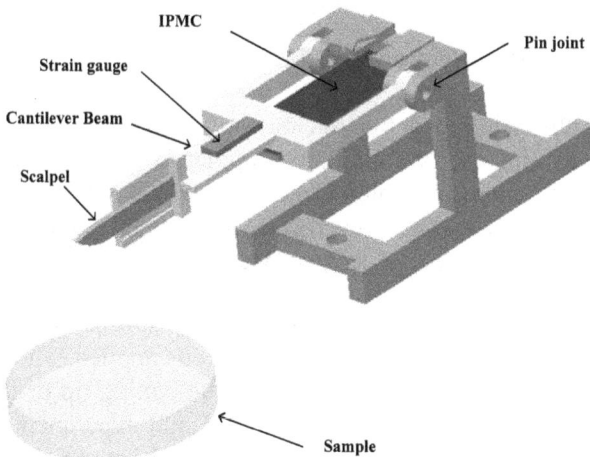

Fig. 9.2. The end effector cutting tool [222, 223].

9.1.2. Mechanical Analysis

Fig. 9.3 shows a finite element analysis (FEA) simulation of the overall deformation when a 10 gf cutting force is generated by the IPMC. 10 gf was used as it was found through preliminary experimentation that this is the maximum output force achievable with this type of IPMC. The scale bar is in millimeters and the major deformation is concentrated (red) at the soft cantilever beam as intended.

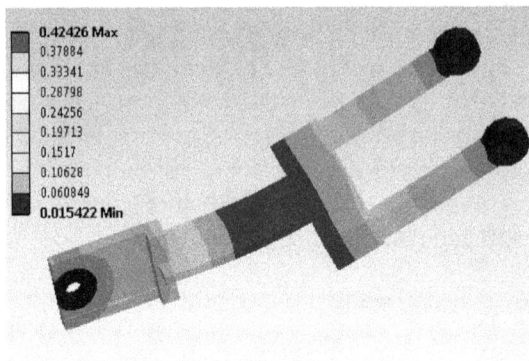

Fig. 9.3. Finite element analysis of the cantilever [222, 223].

From the FEA in Fig. 9.3 the location with the highest stress will be the location where the strain gauge will be located.

9.2.   Integrated Sensor Design

Measuring the bending deflection on the cantilever beam instead of directly measuring the force is much simpler as a strain gauge can be used instead of a more complex force sensor. This will still give a high resolution feedback without the need of complex circuitry or large external sensors such as load cells or laser sensors [216].

A 5 mm foil strain gauge is adhered on to the soft cantilever beam to measure the bending deflection when a force is applied. The deflection should be related to the reaction force at the cutting tip of the scalpel and this will be verified through simulation and experiments. Fig. 9.4 shows the end effector with IPMC actuator and the strain gauge.

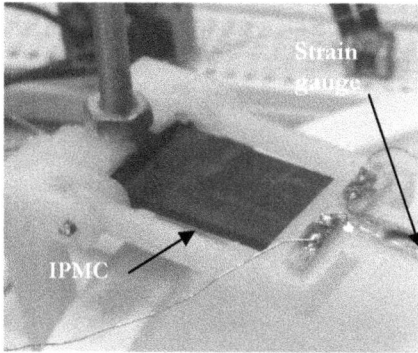

Fig. 9.4. End effector with the IPMC and strain gauge [222].

According to Euler-Bernoulli equation, the displacement on the position of the strain gauge changing with reaction force is defined as below:

$$\delta_S = \frac{P_r}{6EI} x_S^2 \left(3l - x_S\right) = \frac{P_r}{K_M} F\left(x_S\right), \qquad (9.1)$$

where δ_S is the displacement at x_S (position of the strain gauge), P_r is the reaction force, K_M is a material constant, $F(x_s)$ is a function related to the position x_S. This equation was used to determine the dimension of the cantilever in order to obtain a sizeable deflection, which can be measured by the strain gauge, under a typical force by the IPMC.

With the strain gauge strongly glued to the soft cantilever, the stain gauge will bend when the cantilever bends causing a change in the resistance of the strain gauge. This change can be easily measured by the circuit in Fig. 9.5, where R_0 is the original resistance of the strain gauge, is ΔR is resistance difference.

With the strain gauge feeding back the force the cutting depth can be controlled since the depth will increase with the cutting force, which will be further explained in detail in the subsequent section.

The strain gauge is measured by a Wheatstone-bridge circuit as in Fig. 9.5, giving,

$$V_S = \frac{G_A}{2}\left(\frac{\Delta R}{R_0}\right)\left(\frac{1}{2 + \frac{\Delta R}{R_0}}\right)V_e, \qquad (9.2)$$

where V_S is output voltage, V_e is gauge excitation voltage, G_A is gain factor of the amplify circuit.

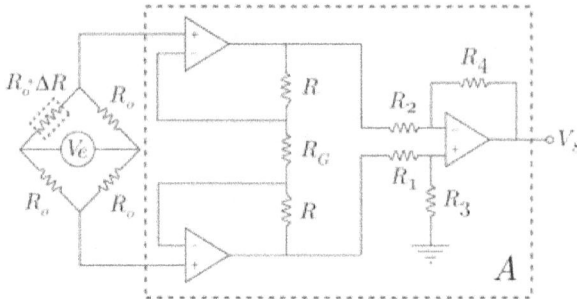

Fig. 9.5. Strain gauge amplifier circuit.

The relationship between resistance difference and strain is

$$\frac{\Delta R}{R_0} = G_f \varepsilon_S, \tag{9.3}$$

where G_f is the gain factor of strain gauge, which is set for a given sensor (can also be determined experimentally) and ε_S is the overall strain of the strain gauge.

The change in the resistance of the strain gauge is usually lower than 2 %, so the equation can be simplified as shown in equations (9.4) and (9.5).

$$V_S \approx \frac{G_A}{4}\left(\frac{\Delta R}{R_0}\right)V_e = \frac{1}{4}G_A G_f \varepsilon_S V_e \tag{9.4}$$

$$G_T = \frac{V_S}{V_e} = \frac{1}{4}G_A G_f \varepsilon_S, \tag{9.5}$$

where the overall gain factor is G_T.

According to the design analysis the strain on the gauge is only approximately 0.1% and hence the required gain factor should be more than 5000.

To calibrate the strain gauge, a load cell (by Sherborne Sensors Ltd., type SS2) is used to measure the cutting force, while the output from the strain gauge amplifier is measured. The amplifier was designed with a gain factor of 5000 which will generate an output voltage of 0.85 to 2.33 V for the corresponding output force of 1.5 to 11.8 gf as shown in Fig. 9.6. The amplifier output voltage versus the force relationship is

$$V = -0.0038F^2 + 0.1922F + 0.5795, \qquad (9.6)$$

where V is the strain gauge amplifier's output voltage and F is the measured force, in gf.

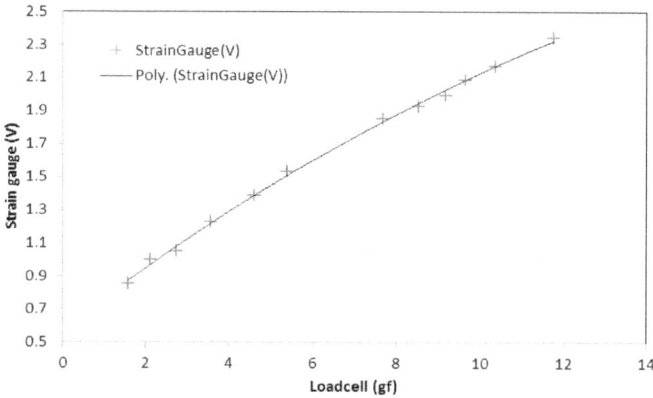

Fig. 9.6. Relationship between strain gauge amplifier's voltage versus the measured force.

9.3. Characterizing the Cutting Depth Versus Force

To test the cutting force required, a replica tissue sample of silicone rubber (ECOFLEX 0030 from SMOOTH-ON) was used, which is slightly harder than human skin. The scalpel was a standard surgery scalpel number 15.

Fig. 9.7 shows the set-up to establish the cutting depth versus the required cutting force.

A 3 mm bolt was screwed in a thread hole on the top of the supporting frame, pushing the skeleton arm downward at same position the IPMC will be actuating from to replicate the IPMC action. The desired force is provided by screwing the bolt up and down. The output force over the tip of scalpel was measured by 'gram force' scale below the silicone rubber.

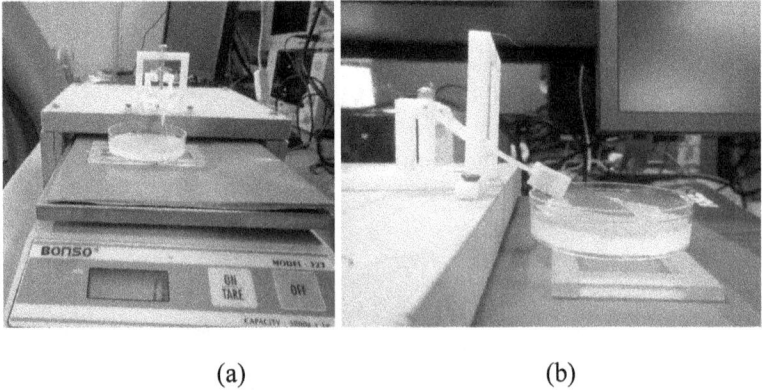

(a)   (b)

Fig. 9.7. Setup for testing the cutting depth [223].

When the required force was reached, the silicon rubber was dragged away from the platform at a constant speed without leaving the surface of the scale, while the scalpel cuts along the silicon rubber to mimic cutting using an actual robotic arm. To see the resulting cutting depth in the silicon a cross section was dissected in normal direction and the depth was assessed under the microscope.

The depth of cutting increased according to the increase of actuation force as illustrated in Fig. 9.8 below, a shot using a microscopic camera. The cut marks in the pictures from Fig. 9.8 (a) to Fig. 9.8 (d) are caused by the scalpel with cutting force from 8 to 14 gf in a 2 gf step.

Fig. 9.9 shows the relationship between the reaction force and cutting depth which can be expressed as a linear relationship of $D = 0.0189$ $F - 0.1132$, where D is the depth, F is the reaction force. By extrapolating the plot it can also be seen that it will take approximately 6 gf to start to pierce the 'skin' of the tissue (in this case the silicon rubber), i.e. to break the surface.

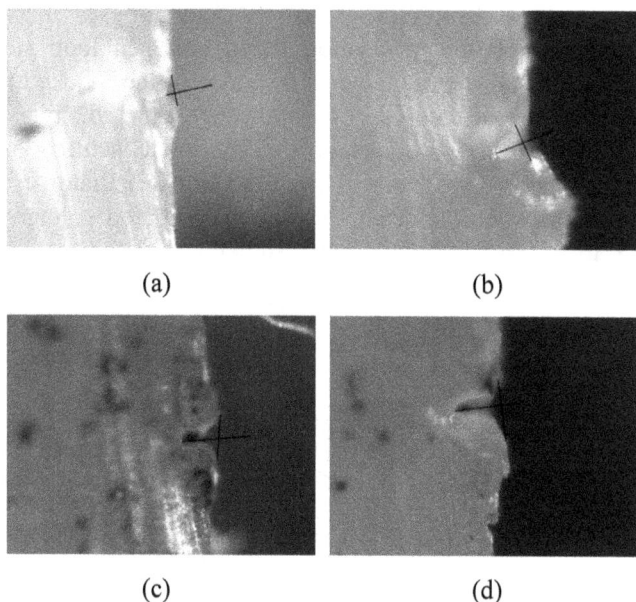

(a) (b)

(c) (d)

Fig 9.8. Cutting depth of (a) 27 μm, (b) 91μm, (c) 114μm and (d) 145 μm into the silicone rubber [222, 223].

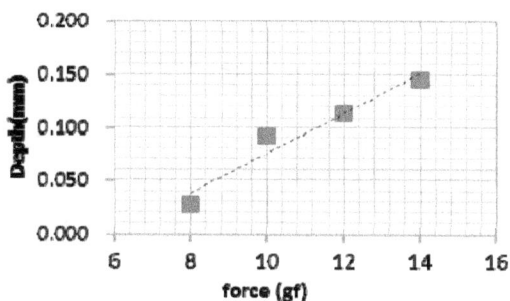

Fig 9.9. Relationship between cutting depth and force [222, 223].

9.4. Control Design

9.4.1. Open Loop Modeling

Previous research in IPMC force control [217, 218] has proved proportional integral controller has acceptable efficiency for the prototype experiments in this research. A model-based method was used to design the force controller, based on open loop step input (0.5 V to 3 V, 0.5 V step increment) response of the blocked force. In this experiment a load cell (by Sherborne Sensors Ltd., type SS2) is used to characterize the force response of the IPMC actuator as shown in Fig. 9.10. The force response data of the IPMC actuator in an open loop is recorded when the IPMC is excited by a step voltage ranging from 0.5 V to 3 V in 0.5 V steps and is shown in Fig. 9.11.

Fig. 9.10. The setup for force control modeling.

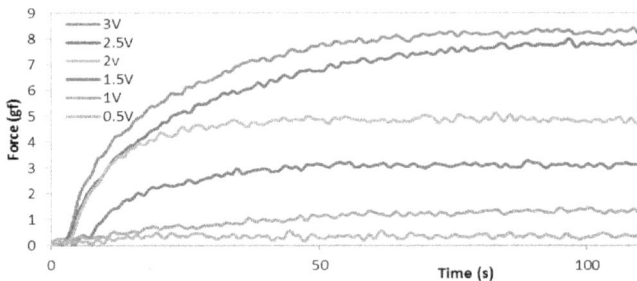

Fig. 9.11. Open loop step stimulation responses [222].

9.4.2. Controller Design

According to the experiments the force response of the IPMC over different input voltage varies significantly. Hence, scheduling different gains with different voltages to the IPMC is necessary. Transfer function models with different input voltages are developed using system identification functions. The best fits are achieved using the transfer function models of 1.0 V for the voltage range of 0.5 V to 1.5 V, while the 2.5 V model can fit the voltage range of 2 V to 3 V. A PI controller was tuned by setting the desired response time 30 seconds and maximum overshoot to 10 %.

Equation (6) is the transfer function model developed for the step response; coefficients vary according to input voltage range.

$$G(s) = K_p \frac{1 + T_z s}{(1 + T_{p1} s)(1 + T_{p2} s)} e^{-T_d s} \qquad (6)$$

The gains are scheduled so that for an input voltage of 1.0 V, K_p= 1.1013, T_{p1}=0.37561, T_{p2}=22.059, T_d=0.1478 and T_z =3.3317. The gains when the input voltage is 2.5 V are K_p= 2.9899, T_{p1}=21.41, T_{p2}=251.01, T_d=0 and T_z =196.46. These gains are used as a GS PI controller for the force feedback control. A PI controller is sufficient in this application.

9.4.3.   Force Feedback Control

The set-up in Fig. 9.12 is a closed-loop system with feedback from the strain gauge and the force sensor used to measure the actual cutting force. Using this set-up the IPMC is controlled by a GS PI controller.

Fig. 9.12. Setup for PI control experiment [222].

Fig. 9.13 shows the various cutting force versus time as provided by the IPMC. It is clear that the simple GS PI controller can be used to provide a range of cutting forces with reasonable accuracy. The actual force is being read by the load cell (blue), while the strain gauge (brown) is used as the feedback to the GS PI controller. From Fig. 9.13 the load cell and strain gauge data track each other and relatively match the requested force (magenta) closely. The output (green) plot is the output voltage from the GS PI controller to the IPMC actuator.

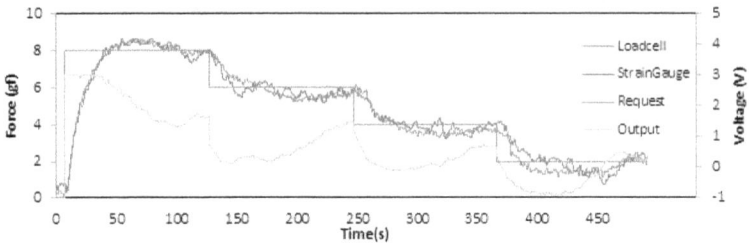

Fig. 9.13. Cutting force controlled by scheduled gain PI controller feedback from the strain gauge [222].

9.4.4. Force Controlled Cutting

Fig. 9.14 (a) shows the experimental set-up to demonstrate the cutting of a replica silicon tissue with a scalpel actuated with an IPMC and force controlled with a gain-schedule PI controller. The arm is lowered until the scalpel touches the silicon and then the GS PI controller will take over. To demonstrate that the cutting force can be controlled with different environmental conditions, the scalpel will cut across a silicon tissue of varying thickness and of constant thickness as well as travelling at varying speeds. Here, the silicone rubber was made in such a manner that the thickness is constant from D to B and increases from 3 mm at C to 9 mm at A in Fig. 9.14 (b).

Fig. 9.15 (a) shows the output voltage from the GS PI controller to the IPMC (dark blue) gradually decreasing when cutting across the silicon tissue that increases from C to A with a reference cutting force of 8 gf (green). The measured force using the strain gauge varies between 8 gf to 10 gf (light blue) as the scalpel travels across from C to A, in an attempt to regulate the cutting force to the required 8 gf. In Fig. 9.15 (b) the travelling speed of the scalpel across D to B is slightly faster (2.1 mm/s) than when cutting across C to A (1.3 mm/s). As the scalpel

speed increases from 1.3 mm/s to 2.1 mm/s at $t = 3$ s, the cutting force
(light blue) increases from the demanded 8 gf (green) and the voltage to
the IPMC (brown) starts to drop in order to reduce the cutting force as
in Fig. 9.15 (b) but only manages to reduce the cutting force to 9 gf at
$t = 10$ s.

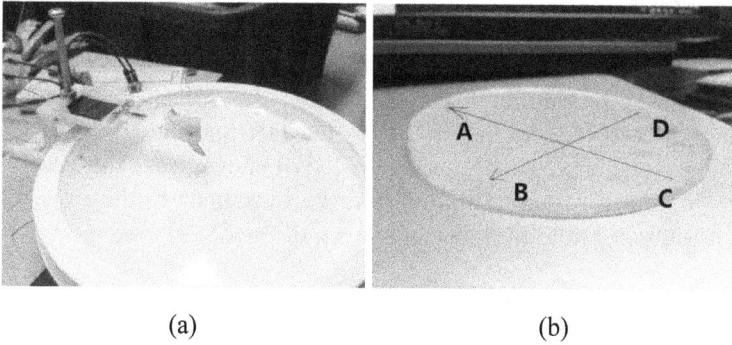

(a)   (b)

Fig.9.14. (a) Experiment setup to demonstrate the force control when cutting
across a silicon tissue, (b) cutting direction with the silicon of constant
thickness (D to B) and of increasing thickness from C to A.

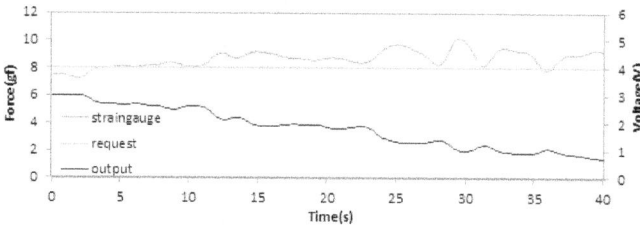

(a)

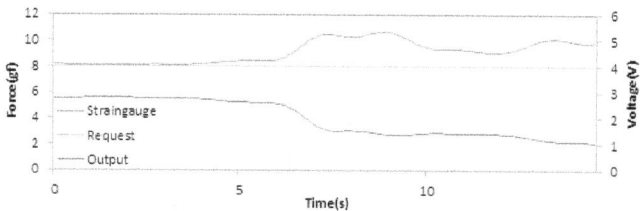

(b)

Fig. 9.15. The output varies with change of the cutting force when cutting (a)
from C to A with increasing thickness, and (b) with a constant thickness from
D to B [222].

215

9.5. Surgical Robotic End Effector Summary

Although it has been shown a simple GS PI controller can to some extent regulate the surgical tool performance it needs to be further improved with a more robust adaptive control. However this design has demonstrated the use of IPMCs in a real world biomedical application.

In the force control experiment it can be seen the force does overshoot from the desired cutting force even when the voltage applied to the IPMC starts to reduce, this is due to the fact that the IPMC is a low frequency response actuator system. It is believed that the compliant nature of the IPMC will limit the force overshoot by less than 2 gf which is the in-built fail-safe mechanism of the IPMC system.

The strain gauge gives sufficiently accurate feedback and is easy to integrate into the surgical device however it is also sensitive to temperature and needs to be compensated in a real application. Also as the strain on the cantilever is small, the gain of the amplifier needs to be high (more than 5000), which introduces unnecessary noise and instabilities that need to be carefully addressed.

Chapter 10

Conclusions

To conclude this book a summary of all the work, which has been undertaken, with respect to the objectives and scope outlined in Chapter 1, is given here. The contributions which have been made to advance the state of art are also presented.

10.1. Research Outcomes

The overall outcome of this research was the development and successful implementation of novel biomedical robotic devices with integrated IPMC actuators.

To achieve this outcome a number of fundamental tasks were completed. These are outlined here.

10.1.1. New Design Based Model of IPMC Actuators

A new comprehensive model for the complete mechanical actuation response of IPMC transducers has been developed. The model is a very useful design tool for developing novel devices, enabling simulations and design optimization before actual systems are built. The model is based on real physical phenomena which makes it scalable for mechanical design as well as relatively simple so that it can also be used for designing and simulating real-time control systems. The model takes into account the effects of external loads and disturbances, which have not been previously achieved.

The model has been used for simulating and verifying the proposed device designs and control schemes and is the foundation building block for developing the biomedical devices in this research. This model represents major progression in IPMC research, enabling a wide variety of previously unachievable research through its increased capabilities and accuracy.

217

10.1.2. Design Novel IPMC Biomedical Devices

With the aid of the newly developed IPMC model and advanced CAD tools six novel devices have been devised, all for distinct biomedical applications. The devices are:

i) A soft compliant stepper motor with a biomimetic fin type actuation mechanism. This has major advantages of biocompatibility, compliance, scalability and hence has potential applications in micromanipulation, microfluidics and position controlled joints.

ii) An artificial muscle finger joint, which has the potential for integration into a hand exoskeleton or human hand prosthesis. The bending type IPMC actuation gives the device the ability to be compact, lightweight and hence operate closely with and even mimic a real hand.

iii) A microfluidic pump for dispensing medical drugs. This device can be used to administer drugs at a more constant rate than traditional intake using pills. The device is mechanically simple, hence scalable, as well as having low power requirements and so the pump is envisaged for miniaturization and implanting in humans.

iv) A cell microtool/gripper which has the ability to handle biological materials safely. The advantages of this device over current techniques are biocompatibility, the ability to operate in an aqueous environment and scalability for operation down to the micro and nano scale.

v) A cell micromanipulator which can achieve high dexterity (greater than 2DOF) with a modular design. This system is inherently compliant and also with the ability to operate accurately in unknown cellular environments has major advantages over traditional cell manipulation techniques.

vi) A surgical robotic end effector actuated by IPMC, which has high compliance as a cutting tool. This tool has an integrated strain gauge to control the cutting force in an attempt to control the cutting depth of tissue.

These systems each have specific functions and must exhibit certain properties to successfully achieve their necessary performance. The devices have been conceived as a result of a thorough design process using the IPMC and mechanical models to simulate and hence confirm their performance.

10.1.3. Advanced Control Methods for Ipmcs

Advances have been made on previous efforts in literature to successfully control the highly nonlinear and time varying properties of IPMCs. To achieve this objective new model free IFT based algorithms have been developed to tune the system performance. The standard IFT algorithm is an adaptive tuning procedure which can account for the time varying nature of the IPMCs themselves. New variations on the IFT algorithms have been developed and implemented customizing the standard tuning algorithm in order to ensure each of the proposed devices exhibit the desired performance characteristics; the finger joint has a nonlinear controller due to its high displacement, the micropump controller is adapted online during normal operation and the microtool/gripper and micromanipulator are robust to external disturbances. These new algorithms have superior performance to controllers which have been previously implemented for controlling IPMCs.

10.1.4. Implement and Test the Biomedical Devices

Finally the proposed devices and their corresponding controllers have been fabricated and then implemented to verify their performance in real life situations. This has validated the entire design process from the IPMC model and mechanical system simulations through to controller performance and device functionality.

By successfully achieving all the above outcomes in this research the feasibility of integrating IPMCs into real world applications as solutions to engineering problems has been proven.

10.2. Contributions to Current State of the Art

Fundamental scientific contributions have been made through novel IPMC modeling and newly improved control algorithms. This underlying scientific theory and investigation has been the major focus of this research and has extended the body of knowledge in this field.

The science has been applied to achieve the overall contribution of developing and implementing new innovative biomedical robotic devices, which will demonstrate the capabilities of non-standard smart material actuators in real world systems.

The scientific and application contributions presented in this research can be split into three parts as follows, where the modeling and control have been uniquely implemented in each of the six devices to achieve the desired system properties.

1. A complete scalable, design based electromechanical IPMC actuator model which allows the design and analysis of any IPMC actuated system. This permits previously unachievable simulations, design optimization and verification of IPMC actuated devices.

2. Improved IFT control algorithms developed to allow more classes of systems to be tuned via this technique:

   i) GS controller for highly nonlinear systems;

   ii) Online adaptive tuning during normal operation;

   iii) Robust control for systems operating in unknown environments with large disturbances.

3. Six novel biomedical robotics devices with integrated smart material IPMC actuators have been designed and discussed in detail:

   i)   Soft compliant stepper motor;

   ii)   Artificial muscle for a finger joint;

   iii)   Microfluidic pump for drug dispensing;

   iv)   Cell microtool/gripper;

   v)   Cell micromanipulator;

   vi)   Robotic surgical cutting tool.

With mechatronics and biomedical research having their foundations over a number of different research disciplines, to contribute to the state of art in these fields a multi-discipline approach has been taken. The contributions in this book span a number of fields as the solution to the research problem being tackled lies in the merger of a number of fundamental research areas.

References

[1]. R. C. Smith, Smart Materials Systems: Model Development, *Society for Industrial and Applied Mathematics*, Philadelphia, 2005.

[2]. F. Capri and E. Smela, Introduction, in Biomedical Applications of Electroactive Polymer Actuators, F. Capri and E. Smela, Eds., *Wiley*, Chichester, UK, 2009.

[3]. K. J. Kim and S. Tadokoro, Electroactive polymers for robotic applications: artificial muscles and sensors, *Springer*, London, 2007.

[4]. M. Shahinpoor, K. J. Kim, and M. Mojarrad, Artificial muscles: applications of advanced polymeric nanocomposites, *Taylor & Francis*, New York, 2007, pp. 119-220.

[5]. M. Shahinpoor and K. J. Kim, Ionic polymer-metal composites: I. Fundamentals, *Smart Materials and Structures*, Vol. 10, 2001, pp. 819-833.

[6]. K. K. Ahn, D. Q. Truong, D. N. C. Nam, J. I. Yoon, and S. Yokota, Position control of ionic polymer metal composite actuator using quantitative feedback theory, *Sensors and Actuators, A: Physical*, Vol. 159, 2010, pp. 204-212.

[7]. K. Mallavarapu and D. J. Leo, Feedback Control of the Bending Response of Ionic Polymer Actuators, *Journal of Intelligent Material Systems and Structures*, Vol. 12, 2001, pp. 143-155.

[8]. N. Bhat and W. J. Kim, Precision force and position control of an ionic polymer metal composite, in *Proceedings of the Institution of Mechanical Engineers, Part I: Journal of Systems and Control Engineering*, Vol. 218, 2004, pp. 421-432.

[9]. S. Zamani and S. Nemat-Nasser, Controlled actuation of Nafion-based ionic polymer-metal composites (IPMCs) with ethylene glycol as solvent., in *Smart Structures and Materials 2004: Electroactive Polymer Actuators and Devices (EAPAD)*, 2004.

[10]. S. Nemat-Nasser and S. Zamani, Modeling of electrochemomechanical response of ionic polymer-metal composites with various solvents, *Journal of Applied Physics*, Vol. 100, 2006, pp. 064310-064310-18.

[11]. M. D. Bennett and D. J. Leo, Ionic liquids as stable solvents for ionic polymer transducers, *Sensors and Actuators, A: Physical*, Vol. 115, 2004, pp. 79-90.

[12]. Y. Bar-Cohen, Electroactive Polymer (EAP) Actuators as Artificial Muscles. Reality, potential and challenges, 2001.

[13]. L. Song-Lin and *et al.*, A helical ionic polymer–metal composite actuator for radius control of biomedical active stents, *Smart Materials and Structures*, Vol. 20, 2011, p. 035008.

[14]. M. Anton, A. Aabloo, A. Punning, and M. Kruusamaa, A mechanical model of a non-uniform ionomeric polymer metal composite actuator, *Smart Materials and Structures*, Vol. 17, 2008, p. 025004.

[15]. M. Shahinpoor and K. J. Kim, Design, development and testing of a multi-fingered heart compression/assist device equipped with IPMC artificial muscles, in *Proceedings of the SPIE - The International Society for Optical Engineering*, 2001, pp. 411-420.

[16]. A. J. McDaid, K. C. Aw, S. Q. Xie, and E. Haemmerle, A conclusive scalable model for the complete actuation response for IPMC transducers, *Smart Materials and Structures*, Vol. 19, 2010, p. 075011.

[17]. S. F. Li, Effect of thickness and length of ion polymer metal composites (IPMC) on its actuation properties, *Advanced Materials Research*, Vol. 197-198, 2011, pp. 401-404.

[18]. M. Yamakita, N. Kamamichi, T. Kozuki, K. Asaka, and L. Zhi-Wei, Control of Biped Walking Robot with IPMC Linear Actuator, in *Proceedings of the IEEE/ASME International Conference on Advanced Intelligent Mechatronics*, 2005, pp. 48-53.

[19]. A. J. McDaid, K. C. Aw, S. Q. Xie, and E. Haemmerle, Gain scheduled control of IPMC actuators with 'model-free' iterative feedback tuning, *Sensors and Actuators A, Physical*, Vol. 164, 2010, pp. 137-147.

[20]. A. Punning, M. Kruusmaa, and A. Aabloo, A self-sensing ion conducting polymer metal composite (IPMC) actuator, *Sensors and Actuators A, Physical*, Vol. 136, 2007, pp. 656-664.

[21]. C. Bonomo, L. Fortuna, P. Giannone, and S. Graziani, A sensor-actuator integrated system based on IPMCs [ionic polymer metal composites], in *Proceedings of the IEEE Sensors*, Vol. 1, 2004, pp. 489-492.

[22]. K. Kruusamäe, P. Brunetto, S. Graziani, L. Fortuna, M. Kodu, R. Jaaniso, A. Punning, and A. Aabloo, Experiments with self-sensing IPMC actuating device, in *Proceedings of the Electroactive Polymer Actuators and Devices (EAPAD)*, SPIE 7642, 2010.

[23]. Y. Nakabo, K. Takagi, T. Mukai, H. Yoshida, and K. Asaka, Bending response of an artificial muscle in high-pressure water environments, in *Proceedings of the SPIE - The International Society for Optical Engineering*, 2005, pp. 388-395.

[24]. X. Ye, Y. Su, S. Guo, and L. Wang, Design and realization of a remote control centimeter-scale robotic fish, in *IEEE/ASME International Conference on Advanced Intelligent Mechatronics (AIM)*, 2008, pp. 25-30.

[25]. Y. Bar-Cohen, S. Leary, A. Yavrouian, K. Oguro, S. Tadokoro, J. Harrison, J. Smith, and J. Su, Challenges to the transition to the practical application of IPMC as artificial-muscle actuators, in *Proceedings of the Materials Research Society Symposium*, 2000, pp. 13-20.

[26]. R. Tiwari and K. J. Kim, Disc-shaped ionic polymer metal composites for use in mechano-electrical applications, *Smart Materials and Structures*, Vol. 19, 2010, p. 065016.

[27]. G.-H. Feng and R.-H. Chen, Fabrication and characterization of arbitrary shaped micro IPMC transducers for accurately controlled biomedical applications, *Sensors and Actuators A: Physical*, Vol. 143, 2008, pp. 34-40.

[28]. G. H. Feng and J. W. Tsai, Development of 3D 4-electrode IPMC actuator with accurate omnidirectional control ability for microendoscopic surgical application, in *Proceedings of the International Solid-State Sensors, Actuators and Microsystems Conference (TRANSDUCERS'09)*, 2009, pp. 2393-2396.

[29]. R. Full and K. Meijer, Metrics of natural muscle, in Electro Active Polymers (EAP) as Artificial Muscles, Reality Potential and Challenges, Y. Bar-Cohen, Ed., *SPIE Press*, 2001, pp. 67-83.

[30]. J. Madden, (2005, 15 June). Actuator Summary. Available Online: http://www.actuatorweb.org/

[31]. I. W. Hunter and S. Lafontaine, A comparison of muscle with artificial actuators, in *Proceedings of the IEEE Solid-State Sensor and Actuator Workshop*, 5th Technical Digest, 1992, pp. 178-185.

[32]. J. D. W. Madden, N. A. Vandesteeg, P. A. Anquetil, P. G. A. Madden, A. Takshi, R. Z. Pytel, S. R. Lafontaine, P. A. Wieringa, and I. W. Hunter, Artificial muscle technology: physical principles and naval prospects, *IEEE Journal of Oceanic Engineering*, Vol. 29, 2004, pp. 706-728.

[33]. Environmental_Robots_Inc. (2011, March 31). The Home of Ionic Polymer Metal Composites (Ipmcs) and Polymeric Nanocomposites. Available, http://www.environmental-robots.com

[34]. C. Zhen, H. Lina, X. Dingyu, X. Xinhe, and L. Yanmei, Modeling and control with hysteresis and creep of ionic polymer-metal composite (IPMC) actuators, in *Proceedings of the Control and Decision Conference (CCDC' 2008)*, Chinese, 2008, pp. 865-870.

[35]. C. Zheng, T. Xiaobo, and M. Shahinpoor, Quasi-Static Positioning of Ionic Polymer-Metal Composite (IPMC) Actuators, in *Proceedings of the IEEE/ASME International Conference on Advanced Intelligent Mechatronics*, 2005, pp. 60-65.

[36]. K. Mallavarapu, K. M. Newbury, and D. J. Leo, Feedback control of the bending response of ionic polymer-metal composite actuators, in *Smart Structures and Materials, Electroactive Polymer Actuators and Devices*, Newport Beach, CA, USA, 2001, pp. 301-310.

[37]. R. C. Richardson, M. C. Levesley, M. D. Brown, J. A. Hawkes, K. Watterson, and P. G. Walker, Control of ionic polymer metal composites, *IEEE/ASME Transactions on Mechatronics*, Vol. 8, 2003, pp. 245-253.

[38]. C. Bonomo, L. Fortuna, P. Giannone, S. Graziani, and S. Strazzeri, A nonlinear model for ionic polymer metal composites as actuators, *Smart Materials and Structures*, Vol. 16, 2007, pp. 1-12.

[39]. B. C. Lavu, M. P. Schoen, and A. Mahajan, Adaptive intelligent control of ionic polymer-metal composites, *Smart Materials and Structures*, Vol. 14, 2005, pp. 466-474.

[40]. J. Brufau-Penella, K. Tsiakmakis, T. Laopoulos, and M. Puig-Vidal, Model reference adaptive control for an ionic polymer metal composite in underwater applications, *Smart Materials and Structures*, Vol. 17, 2008.

[41]. S. Lee, H. C. Park, and K. J. Kim, Equivalent modeling for ionic polymer - Metal composite actuators based on beam theories, *Smart Materials and Structures*, Vol. 14, 2005, pp. 1363-1368.

[42]. T. Yih and I. Talpasanu, Introduction, in Micro and Nano Manipulations for Biomedical Applications, T. Yih and I. Talpasanu, Eds., *Artech House, Inc.*, Boston, 2008.

[43]. M. Shahinpoor, Implantable Heart-Assist and Compression Devices Employing an Active Network of Electrically-Controllable Ionic Polymer-Metal Nanocomposites, in Biomedical Applications of Electroactive Polymer Actuators, F. Capri and E. Smela, Eds., *Wiley,* Chichester, UK, 2009.

[44]. A. J. McDaid, K. C. Aw, S. Q. Xie, and E. Haemmerle, Development of a 2DOF Micromanipulation System with IPMC Actuators, in *Proceedings of the IEEE/ASME International Conference on Advanced Intelligent Mechatronics,* AIM, Budapest, Hungary, 2011.

[45]. F. Ionescu, K. Kostadinov, I. Talpasanu, D. Arotaritei, and G. Constantin, Design, Analysis, Modelling, Simulation and Control of Microscale and Nanoscale Cell Manipulations, in Micro and Nano Manipulations for Biomedical Applications, T. Yih and I. Talpasanu, Eds., *Artech House, Inc.,* Boston, 2008.

[46]. B. K. Fang, M. S. Ju, and C. C. K. Lin, A new approach to develop ionic polymer-metal composites (IPMC) actuator: Fabrication and control for active catheter systems, *Sensors and Actuators, A: Physical,* Vol. 137, 2007, pp. 321-329.

[47]. H. H. Lin, B. K. Fang, M. S. Ju, and C. C. K. Lin, Control of ionic polymer-metal composites for active catheter systems via linear parameter-varying approach, *Journal of Intelligent Material Systems and Structures,* Vol. 20, 2009, pp. 273-282.

[48]. G. M. Spinks and G. G. Wallace, Actuated Pins for Braille Displays, in Biomedical Applications of Electroactive Polymer Actuators, F. Capri and E. Smela, Eds., *Wiley,* Chichester, UK, 2009.

[49]. P. Chiarelli and P. Ragni, Thermally Driven Hydrogel Actuator for Controllable Flow Rate Pump in Long Term Drug Delivery, in Biomedical Applications of Electroactive Polymer Actuators, F. Capri and E. Smela, Eds., *Wiley,* Chichester, UK, 2009.

[50]. A. O. Saeed, J. P. Magnusson, B. Twaites, and C. Alexander, Stimuli-Responsive and 'Active' Polymers in Drud Delivery, in Biomedical Applications of Electroactive Polymer Actuators, F. Capri and E. Smela, Eds., *Wiley,* Chichester, UK, 2009.

[51]. A. J. McDaid, K. C. Aw, and S. Q. Xie, Modeling and Control of Ionic Polymer-Metal Composite (IPMC) Actuators for Mechatronics Applications, in Mechatronics, J. P. Davim, Ed., *ISTE-Wiley*, 2011.

[52]. J. K. Kwang and S. Mohsen, Ionic polymer–metal composites: II. Manufacturing techniques, *Smart Materials and Structures*, Vol. 12, 2003, p. 65.

[53]. M. Bennett, Manufacture and Characterization of Ionic Polymer Transducers Employing Non-Precious Metal Electrodes, M. S. M. S. Thesis, *Virginia Tech*, 2002.

[54]. K. Bian, K. Xiong, Q. Chen, G. Liu, and B. Wang, Manufacture and actuating characteristic of ionic polymer metal composites with silver electrodes, *Cailiao Yanjiu Xuebao, Chinese Journal of Materials Research*, Vol. 24, 2010, pp. 520-524.

[55]. N. Jin, B. Wang, K. Bian, Q. Chen, and K. Xiong, Performance of ionic polymer-metal composite (IPMC) with different surface roughening methods, *Frontiers of Mechanical Engineering in China*, Vol. 4, 2009, pp. 430-435.

[56]. K. S. Lee, B. J. Jeon, and S. W. Cha, Performance enhancement of an ionic polymer metal composite actuator using a microcellular foaming process, *Smart Materials and Structures*, Vol. 19, 2010.

[57]. I. S. Park, C. Bae, T. S. Jo, J. Truong, S. M. Kim, K. J. Kim, W. Yim, and J. S. Lee, Sulfonated polyamide based IPMCs, 2009.

[58]. R. Tiwari and K. J. Kim, Effect of metal diffusion on mechanoelectric property of ionic polymer-metal composite, *Applied Physics Letters*, Vol. 97, 2010.

[59]. R. Tiwari and K. J. Kim, Disc-shaped ionic polymer metal composites for use in mechano-electrical applications, *Smart Materials and Structures*, Vol. 19, 2010.

[60]. K. Sadeghipour, R. Salomon, and S. Neogi, Development of a novel electrochemically active membrane and `smart' material based vibration sensor/damper, *Smart Materials and Structures*, Vol. 1, 1992, pp. 172-179.

[61]. C. Bonomo, L. Fortuna, P. Giannone, S. Graziani, and S. Strazzeri, A model for ionic polymer metal composites as sensors, *Smart Materials and Structures*, Vol. 15, 2006, pp. 749-758.

[62]. Z. Chen, X. Tan, A. Will, and C. Ziel, A dynamic model for ionic polymer-metal composite sensors, *Smart Materials and Structures*, Vol. 16, 2007, pp. 1477-1488.

[63]. A. Punning, M. Kruusmaa, and A. Aabloo, Surface resistance experiments with IPMC sensors and actuators, *Sensors and Actuators, A: Physical,* Vol. 133, 2007, pp. 200-209.

[64]. X. J. Chew, A. V. D. Hurk, and K. C. Aw, Characterisation of ionic polymer metallic composites as sensors in robotic finger joints, *International Journal of Biomechatronics and Biomedical Robotics*, Vol. 2, 2009, pp. 37-43.

[65]. K. Park, M. K. Yoon, S. Lee, J. Choi, and M. Thubrikar, Effects of electrode degradation and solvent evaporation on the performance of ionic-polymer-metal composite sensors, *Smart Materials and Structures,* Vol. 19, 2010.

[66]. C. Bonomo, L. Fortuna, S. Graziani, M. L. Rosa, D. Nicolosi, and G. Sicurella, Towards biocompatible sensing devices: An IPMC based artificial vestibular system, 2008, pp. 85-90.

[67]. P. Brunetto, L. Fortuna, P. Giannone, S. Graziani, and F. Pagano, A small scale viscometer based on an IPMC actuator and an IPMC sensor, 2010, pp. 585-589.

[68]. B. G. Ko, H. C. Kwon, and S. Lee, A self-sensing method for IPMC actuator, *Advances in Science and Technology*, Vol. 56, 2008, pp. 111-115.

[69]. I. S. Park, R. Tiwari, and K. J. Kim, Sprayed sensor using IPMC paint, 2008, pp. 59-64.

[70]. R. P. Hamlen, C. E. Kent, and S. N. Shafer, Electrolytically activated contractile polymer, *Nature*, Vol. 206, 1965, pp. 1149-1150.

[71]. K. Oguro, Y. Kawagami, and H. Takenaka, Bending of an Ion-Conducting Polymer Film-Electrode Composite by an Electric Stimulus at Low Voltage, *Journal of Micromachine Society*, Vol. 5, 1992, pp. 27-30.

[72]. M. Shahinpoor, Conceptual design, kinematics and dynamics of swimming robotic structures using ionic polymeric gel muscles, *Smart Materials and Structures*, Vol. 1, 1992, pp. 91-94.

[73]. R. Kanno, A. Kurata, M. Hattori, S. Tadokoro, and T. Takamori, Characteristics and Modeling of ICPF Actuator, in *Proceedings of the Japan-USA Symposium on Flexible Automation*, 1994, pp. 691-698.

[74]. D. J. Leo, K. Farinholt, and T. Wallmersperger, Computational models of ionic transport and electromechanical transduction in ionomeric polymer transducers, in *Proceedings of the SPIE-Electroactive Polymers and Devices (EAPAD)*, 2005, pp. 5759-24.

[75]. G. D. Bufalo, L. Placidi, and M. Porfiri, A mixture theory framework for modeling the mechanical actuation of ionic polymer metal composites, *Smart Materials and Structures,* Vol. 17, 2008.

[76]. K. Yagasaki and H. Tamagawa, Experimental estimate of viscoelastic properties for ionic polymer-metal composites, *Physical Review E,* Vol. 70, 2004, p. 052801.

[77]. K. Mallavarapu, Feedback Control of Ionic Polymer Actuators, Master of Science, *Virginia Polytechnic Institute and State University,* Virginia, 2001.

[78]. Y. Xiao and K. Bhattacharya, Modeling electromechanical properties of ionic polymers, in *Proceedings of the SPIE - The International Society for Optical Engineering,* 2001, pp. 292-300.

[79]. S. Kang, J. Shin, S. J. Kim, H. J. Kim, and Y. H. Kim, Robust control of ionic polymer-metal composites, *Smart Materials and Structures,* Vol. 16, 2007, pp. 2457-2463.

[80]. S. Nemat-Nasser and J. Y. Li, Electromechanical response of ionic polymer-metal composites, *Journal of Applied Physics,* Vol. 87, 2000, pp. 3321-3331.

[81]. D. Q. Truong, K. K. Ahn, D. N. C. Nam, and J. I. Yoon, Estimation of bending behavior of an ionic polymer metal composite actuator using a nonlinear black-box model, 2010, pp. 438-442.

[82]. R. Caponetto, G. Dongola, L. Fortuna, S. Graziani, and S. Strazzeri, A Fractional Model for IPMC Actuators, in *Proceedings of the IEEE Instrumentation and Measurement Technology Conference (IMTC'08),* 2008, pp. 2103-2107.

[83]. D. K. Biswal, D. Bandopadhya, and S. K. Dwivedy, Evaluation of dehydration loss and investigation of its effect on bending response of segmented IPMC actuators, *International Journal of Smart and Nano Materials,* Vol. 1, 2010, pp. 187 - 200.

[84]. L. M. Weiland and J. L. D, Electrostatic analysis of cluster response to electrical and mechanical loading in ionic polymers with cluster morphology, *Smart Materials and Structures*, Vol. 13, 2004, p. 323.

[85]. S. Tadokoro, S. Yamagami, T. Takamori, and K. Oguro, Modeling of Nafion-Pt composite actuators (ICPF) by ionic motion, *Smart Structures and Materials 2000: Electroactive Polymer Actuators and Devices (EAPAD)*, Newport Beach, CA, USA, 2000, pp. 92-102.

[86]. M. Shahinpoor, Electromechanics of ionoelastic beams as electrically controllable artificial muscles, *Smart Structures and Materials 1999: Electroactive Polymer Actuators and Devices,* Newport Beach, CA, USA, 1999, pp. 109-121.

[87]. M. Shahinpoor, Micro-electro-mechanics of ionic polymeric gels as electrically controllable artificial muscles, *Journal of Intelligent Material Systems and Structures*, Vol. 6, 1995, pp. 307-314.

[88]. K. Asaka and K. Oguro, Bending of polyelectrolyte membrane platinum composites by electric stimuli: Part II. Response kinetics, *Journal of Electroanalytical Chemistry*, Vol. 480, 2000, pp. 186-198.

[89]. L. M. Weiland and D. J. Leo, Computational analysis of ionic polymer cluster energetics, *Journal of Applied Physics*, Vol. 97, 2005, pp. 013541-10.

[90]. W. A. Lughmani, J. Y. Jho, J. Y. Lee, and K. Rhee, Modeling of bending behavior of IPMC beams using concentrated ion boundary layer, *International Journal of Precision Engineering and Manufacturing,* Vol. 10, 2009, pp. 131-139.

[91]. M. Shahinpoor, Modelling of large deflection of IPMC plates, in *Proceedings of the ASME Dynamic Systems and Control Conference,* 2010, pp. 461-468.

[92]. P. J. C. Branco and J. A. Dente, Derivation of a continuum model and its electric equivalent-circuit representation for ionic polymer-metal composite (IPMC) electromechanics, *Smart Materials and Structures,* Vol. 15, 2006, pp. 378-392.

[93]. R. Kanno, S. Tadokoro, T. Takamori, M. Hattori, and K. Oguro, Linear approximate dynamic model of ICPF (ionic conducting polymer gel film) actuator, in *Proceedings of the IEEE International Conference on Robotics and Automation,* Vol. 1, 1996, pp. 219-225.

[94]. R. Kanno, S. Tadokoro, T. Takamori, and K. Oguro, 3-dimensional dynamic model of ionic conducting polymer gel film (ICPF) actuator, in *Proceedings of the IEEE International Conference on Systems, Man, and Cybernetics*, 1996, Vol. 3, pp. 2179-2184.

[95]. P. G. deGennes, K. Okumura, M. Shahinpoor, and K. J. Kim, Mechanoelectric effects in ionic gels, *Europhysics Letters*, Vol. 50, 15 May 2000.

[96]. J. W. Paquette, K. J. Kim, J.-D. Nam, and Y. S. Tak, An Equivalent Circuit Model for Ionic Polymer-Metal Composites and their Performance Improvement by a Clay-Based Polymer Nano-Composite Technique, *Journal of Intelligent Material Systems and Structures*, Vol. 14, October 1, 2003, pp. 633-642.

[97]. K. M. Newbury and D. J. Leo, Electromechanical Modeling and Characterization of Ionic Polymer Benders, *Journal of Intelligent Material Systems and Structures*, Vol. 13, January 1, 2002, pp. 51-60.

[98]. K. M. Newbury and D. J. Leo, Linear Electromechanical Model of Ionic Polymer Transducers. Part I: Model Development, *Journal of Intelligent Material Systems and Structures*, Vol. 14, June 1, 2003, pp. 333-342.

[99]. K. M. Newbury and D. J. Leo, Linear Electromechanical Model of Ionic Polymer Transducers, Part II: Experimental Validation, *Journal of Intelligent Material Systems and Structures*, Vol. 14 June 1, 2003, pp. 343-357.

[100]. K. M. Newbury, Characterization, Modeling, and Control of Ionic Polymer Transducers, PhD Thesis, Mechanical Engineering, *Virginia Polytechnic Institute and State University*, Virginia, 2002.

[101]. K. Yun, A Novel Three-Finger IPMC Gripper for Microscale Applications, PhD Thesis, Mechanical Engineering, *Texas A&M University*, 2006.

[102]. C. S. Kothera, Characterization, Modeling, and Control of the Nonlinear Actuation Response of Ionic Polymer Transducers, PhD Thesis, Mechanical Engineering, *Virginia Polytechnic Institute and State University*, 2005.

[103]. D. Liu, Design and Control of an IPMC Actuated Single Degree of Freedom Rotary Joint, Master's Thesis, Mechatronics Engineering, *The University of Auckland*, NZ, Auckland, 2010.

[104]. K. Yun and W. J. Kim, Microscale position control of an electroactive polymer using an anti-windup scheme, *Smart Materials and Structures*, Vol. 15, 2006, pp. 924-930.

[105]. L.-N. H. Zhen Chen, Ding-Yu Xue, Xin-He Xu, Yan-Mei Liu, Real-time compensation control for hysteresis and creep in IPMC actuators, *International Journal of Modelling, Identification and Control (IJMIC)*, Vol. 12, 2011.

[106]. Y. O. Minoru Sasaki, Hirohisa Tamagawa, Satoshi Ito, Two-Degree-of-Freedom Control of an Ionic Polymer-Metal Composite Actuator, *Materials Science Forum*, Vol. 670, 2010.

[107]. M. Sasaki, Y. Onouchi, T. Ozeki, H. Tamagawa, and S. Ito, Feedforward control of an Ionic Polymer-Metal Composite actuator, *International Journal of Applied Electromagnetics and Mechanics*, Vol. 33, 2010, pp. 875-881.

[108]. S. Yingfeng and K. L. Kam, Frequency-weighted feedforward control for dynamic compensation in ionic polymer–metal composite actuators, *Smart Materials and Structures*, Vol. 18, 2009, p. 125016.

[109]. K. Tsiakmakis, J. Brufau, M. Puig-Vidal, and T. Laopoulos, Modeling IPMC Actuators for Model Reference Motion Control, in *Proceedings of the IEEE Instrumentation and Measurement Technology Conference (IMTC'08)*, 2008, pp. 1168-1173.

[110]. H. Khadivi, B. S. Aghazadeh, and C. Lucas, Fuzzy control of ionic polymer-metal composites, in *Proceedings of the IEEE Annual International Conference of the Engineering in Medicine and Biology*, 2007, pp. 4198-4201.

[111]. Z. Chen and X. Tan, A control-oriented and physics-based model for ionic polymer-metal composite actuators, *IEEE/ASME Transactions on Mechatronics*, Vol. 13, 2008, pp. 519-529.

[112]. A. Wang, M. Deng, and D. Wang, Operator-Based Robust Nonlinear Control for Ionic Polymer Metal Composite with Uncertainties and Hysteresis, in Intelligent Robotics and Applications. Vol. 6424, H. Liu, *et al.*, Eds., *Springer*, Berlin / Heidelberg, 2010, pp. 135-146.

[113]. S. Sano, K. Takagi, S. Sato, S. Hirayama, N. Uchiyama, and K. Asaka, Robust PID force control of IPMC actuators, 2010.

[114]. T. T. Nguyen, Y. S. Yang, and I. K. Oh, Position control of ionic polymer metal composite actuator based on neuro-fuzzy system, in

Proceedings of the 2nd International Conference on Smart Materials and Nanotechnology in Engineering, SPIE 7493, 2009.

[115]. A. Shariati, A. Meghdari, and P. Shariati, Intelligent control of an IPMC actuated manipulator using emotional learning-based controller, in *Proceedings of the SPIE - The International Society for Optical Engineering*, 2008, p. 70291J.

[116]. H. M. La and W. Sheng, Robust adaptive control with leakage modification for a nonlinear model of ionic polymer metal composites (IPMC), in *Proceedings of the IEEE International Conference on Robotics and Biomimetics (ROBIO' 2008)*, pp. 1783-1788.

[117]. L. Myung-Joon, J. Sung-Hee, L. Sukmin, A. M.-S. M. Mu-Seong Mun, and A. I. M. Inhyuk Moon, Control of IPMC-based Artificial Muscle for Myoelectric Hand Prosthesis, in *Proceedings of the 1st IEEE/RAS-EMBS International Conference on Biomedical Robotics and Biomechatronics (BioRob'06)*, 2006, pp. 1172-1177.

[118]. S. Manley, Performance and Feasibility of a Miniature Single Degree-of-Freedom Rotary Mechanism with Integrated IPMC Actuator, Master's Thesis, Mechatronics Engineering, *The University of Auckland*, NZ, Auckland, 2009.

[119]. S. Manley, A. McDaid, K. Aw, E. Haemmerle, and S. Xie, Experimental Performance and Feasibility of a Miniature Single-Degree-Of-Freedom Rotary Joint with Integrated IPMC Actuator, *Electroactive Polymers and Devices*, San Diego, USA, 2009.

[120]. A. Van Den Hurk, A rotary joint sensor using ionic polymer metallic composite, Bachelor of Engineering, Mechanical Engineering, *The University of Auckland*, Auckland, 2009.

[121]. A. van den Hurk, X. J. Chew, K. C. Aw, and S. Q. Xie, A rotary joint sensor using ionic polymer metallic composite, in *Proceedings of the Second International Conference on Smart Materials and Nanotechnology in Engineering*, Weihai, China, 2009, pp. 74932K-8.

[122]. P. Aravinthan, N. Gopala Krishnan, P. A. Srinivas, and N. Vigneswaran, Design, development and implementation of neurologically controlled prosthetic limb capable of performing rotational movement, in *Proceedings of the International Conference on Emerging Trends in Robotics and Communication Technologies (INTERACT)*, 2010, pp. 241-244.

[123]. K. Asaka and K. Oguro, Active Microcatheter and Biomedical Soft Devices Based on IPMC Actuators, in Biomedical Applications of Electroactive Polymer Actuators, F. Capri and E. Smela, Eds., *Wiley,* Chichester, UK, 2009.

[124]. G. Shuxiang, T. Fukuda, K. Kosuge, F. A. A. F. Arai, K. A. O. K. Oguro, and M. A. N. M. Negoro, Micro catheter system with active guide wire, in *Proceedings of the IEEE International Conference on Robotics and Automation,* Vol. 1, 1995, pp. 79-84.

[125]. G. H. Feng and J. W. Tsai, Micromachined optical fiber enclosed 4-electrode IPMC actuator with multidirectional control ability for biomedical application, *Biomedical Microdevices,* pp. 1-9, 2010.

[126]. J. Santos, B. Lopes, and P. J. C. Branco, Ionic polymer-metal composite material as a diaphragm for micropump devices, *Sensors and Actuators, A: Physical,* Vol. 161, 2010, pp. 225-233.

[127]. I.-S. Park, S. Vohnout, M. Banister, S. Lee, S.-M. Kim, and K. Kim, IPMC Assisted Infusion Micropumps, in Biomedical Applications of Electroactive Polymer Actuators, F. Capri and E. Smela, Eds., *Wiley,* Chichester, UK, 2009.

[128]. M. Konyo and S. Tadokoro, IPMC Based Tactile Displays for Pressure and Texture Presentation on a Human Finger, in Biomedical Applications of Electroactive Polymer Actuators, F. Capri and E. Smela, Eds., *Wiley,* Chichester, UK, 2009.

[129]. E. Mbemmo, Z. Chen, S. Shatara, and X. Tan, Modeling of biomimetic robotic fish propelled by an ionic polymer-metal composite actuator, in *Proceedings of the IEEE International Conference on Robotics and Automation,* 2008, pp. 689-694.

[130]. 31 March). EAMEX Corporation, Japan. Available Online: http://www.eamex.co.jp/index_e.html

[131]. A. Hunt, A. Punning, M. Anton, A. Aabloo, and M. Kruusmaa, A multilink manipulator with IPMC joints, in *Proceedings of the SPIE - The International Society for Optical Engineering,* 2008.

[132]. N. Kamamichi, M. Yamakita, K. Asaka, and L. Zhi-Wei, A snake-like swimming robot using IPMC actuator/sensor, in *Proceedings of the IEEE International Conference on Robotics and Automation (ICRA'06),* 2006, pp. 1812-1817.

[133]. S. Guo, L. Shi, X. Ye, and L. Li, A new jellyfish type of underwater microrobot, in *Proceedings of the IEEE International Conference on Mechatronics and Automation (ICMA '07)*, 2007, pp. 509-514.

[134]. S. Mukherjee and R. Ganguli, Ionic polymer metal composite flapping actuator mimicking Dragonflies, *Computers, Materials and Continua*, Vol. 19, pp. 105-133, 2010.

[135]. H. Ji Ping, L. Min Zhou, and M. Tao, The design and control of Amoeba-like robot, in *Proceedings of the International Conference on Computer Application and System Modeling (ICCASM)*, 2010, pp. V1-88-V1-91.

[136]. P. Arena, C. Bonomo, L. Fortuna, M. Frasca, and S. Graziani, Design and Control of an IPMC Wormlike Robot, *IEEE Transactions on Systems, Man, and Cybernetics*, Part B: Cybernetics, Vol. 36, 2006, pp. 1044-1052.

[137]. D. Bandopadhya, D. K. Bhogadi, B. Bhattacharya, and A. A. D. A. Dutta, Active Vibration Suppression of a Flexible Link Using Ionic Polymer Metal Composite, in *Proceedings of the IEEE Conference on Robotics, Automation and Mechatronics*, 2006, pp. 1-6.

[138]. D. Bandopadhya, B. Bhattacharya, and A. Dutta, Active Vibration Control Strategy for a Single-Link Flexible Manipulator Using Ionic Polymer Metal Composite, *Journal of Intelligent Material Systems and Structures*, Vol. 19, 2008, pp. 487-496.

[139]. M. Shahinpoor and K. J. Kim, Ionic polymer-metal composites: IV. Industrial and medical applications, *Smart Materials and Structures*, Vol. 14, pp. 197-214, 2005.

[140]. M. Anton, M. Kruusmaa, A. Aabloo, and A. Punning, Validating Usability of Ionomeric Polymer-Metal Composite Actuators for Real World Applications, in *Proceedings of the IEEE/RSJ International Conference on Intelligent Robots and Systems*, 2006, pp. 5441-5446.

[141]. A. Hunt, C. Zheng, T. Xiaobo, and M. Kruusmaa, Control of an inverted pendulum using an Ionic Polymer-Metal Composite actuator, in *Proceedings of the IEEE/ASME International Conference on Advanced Intelligent Mechatronics (AIM)*, 2010, pp. 163-168.

[142]. C. Bonomo and *et al.*, Tridimensional ionic polymer metal composites: optimization of the manufacturing techniques, *Smart Materials and Structures*, Vol. 19, 2010, p. 055002.

[143]. Y. Bar-Cohen, T. Xue, M. Shahinpoor, J. Simpson, and J. Smith, Flexible, low-mass robotic arm actuated by electroactive polymers and operated equivalently to human arm and hand, in *Proceedings of the ASCE Specialty Conference on Robotics for Challenging Environments*, 1998, pp. 15-21.

[144]. K. S. Yun, A novel three-finger IPMC gripper for microscale applications, PhD Dissertation, Mechanical Engineering, *Texas A&M University*, 2006.

[145]. U. Dcole, R. Lumia, and M. Shahinpoor, Grasping flexible objects using artificial muscle microgrippers, in *Proceedings of the 6th Biannual World Automation Congress in Robotics: Trends, Principles, and Applications*, 2004, pp. 191-196.

[146]. R. Lumia and M. Shahinpoor, IPMC microgripper research and development, *Journal of Physics, Conference Series*, Vol. 127, 2008, p. 012002.

[147]. L. Joon Soo, S. Gutta, Y. Woosoon, and K. J. Kim, Preliminary study of wireless actuation and control of IPMC actuator, in *Proceedings of the IEEE/ASME International Conference on Advanced Intelligent Mechatronics (AIM)*, 2010, pp. 157-162.

[148]. Y. Bar-Cohen, S. Leary, A. Yavrouian, K. Oguro, S. Tadokoro, J. Harrison, J. Smith, and J. Su, Challenges to the application of IPMC as actuators of planetary mechanisms, In *Proceedings of the SPIE - The International Society for Optical Engineering*, Vol. 3987, 2000, pp. 140-146.

[149]. A. J. McDaid, K. Aw, S. Q. Xie, and E. Haemmerle, A Nonlinear Scalable Model for Designing Ionic Polymer-Metal Composite Actuator Systems, in *Proceedings of the 2nd International Conference on Smart Materials and Nanotechnology in Engineering*, WeiHai, China, 2009.

[150]. C. S. Kothera, D. J. Leo, and S. L. Lacy, Characterization and Modeling of the Nonlinear Response of Ionic Polymer Actuators, *Journal of Vibration and Control*, Vol. 14, 2008, pp. 1151-1173.

[151]. Z. Chen, D. R. Hedgepeth, and X. Tan, A nonlinear, control-oriented model for ionic polymer-metal composite actuators, *Smart Materials and Structures*, Vol. 18, 2009.

[152]. M. Porfiri, Charge dynamics in ionic polymer metal composites, *Journal of Applied Physics*, Vol. 104, 2008.

[153]. R. C. Hibbeler, Mechanics of Materials, *Prentice Hall,* Singapore, 2004.

[154]. M. A. Trindade, A. Benjeddou, and R. Ohayon, Modeling of Frequency-Dependent Viscoelastic Materials for Active-Passive Vibration Damping, *Journal of Vibration and Acoustics,* Vol. 122, 2000, pp. 169-174.

[155]. J. C. Lagarias, J. A. Reeds, M. H. Wright, and P. E. Wright, Convergence Properties of the Nelder-Mead Simplex Method in Low Dimensions, *SIAM Journal of Optimization,* Vol. 9, 1998, pp. 112-147.

[156]. A. J. McDaid, K. C. Aw, K. Patel, S. Q. Xie, and E. Haemmerle, Development of an ionic polymer–metal composite stepper motor using a novel actuator model, *International Journal of Smart and Nano Materials,* Vol. 1, 2010, pp. 261-277.

[157]. K. Patel, Design, Simulation and Performance Evaluation of a Miniature IPMC Actuated Rotary Device, Master's Thesis, Mechatronics Engineering, *The University of Auckland,* NZ, Auckland, 2010.

[158]. K. S. Narendra and D. N. Streeter, An adaptive procedure for controlling undefined linear processes, *IEEE Transactions on Automatic Control,* 1964, pp. 545-548.

[159]. H. Hjalmarsson, S. Gunnarsson, and M. Gevers, A convergent iterative restricted complexity control design scheme, in *Proceedings of the 33rd IEEE Conference on Decision and Control,* 1994, pp. 1735-1740.

[160]. A. E. Graham, A. J. Young, and S. Q. Xie, Rapid tuning of controllers by IFT for profile cutting machines, *Mechatronics,* Vol. 17, 2007, pp. 121-128.

[161]. S. Kissling, P. Blanc, P. Myszkorowski, and I. Vaclavik, Application of iterative feedback tuning (IFT) to speed and position control of a servo drive, *Control Engineering Practice,* Vol. 17, 2009, pp. 834-840.

[162]. H. Hjalmarsson, M. Gevers, S. Gunnarsson, and O. Lequin, Iterative feedback tuning-theory and applications, *IEEE Control Systems Magazine,* Vol. 18, 1998, pp. 26-41.

[163]. A. Tay, W. Khuen Ho, J. Deng, and B. Keng Lok, Control of photoresist film thickness: Iterative feedback tuning approach, *Computers & Chemical Engineering,* Vol. 30, 2006, pp. 572-579.

[164]. B. Codrons, F. De Bruyne, M. De Wan, and M. Gevers, Iterative feedback tuning of a nonlinear controller for an inverted pendulum with a flexible transmission, in *Proceedings of the IEEE International Conference on Control Applications,* Vol. 2, 1998, pp. 1281-1285.

[165]. A. Karimi, L. Miskovic, and D. Bonvin, Iterative correlation-based controller tuning with application to a magnetic suspension system, *Control Engineering Practice,* Vol. 11, 2003, pp. 1069-1078.

[166]. T. Meurers, S. M. Veres, and A. C. H. Tan, Model-free frequency domain iterative active sound and vibration control, *Control Engineering Practice,* Vol. 11, 2003, pp. 1049-1059.

[167]. H. Hjalmarsson, Iterative feedback tuning - an overview, International *Journal of Adaptive Control and Signal Processing,* Vol. 16, 2002, pp. 373-395.

[168]. A. I. Talkin, Adaptive servo tracking, *IRE Transactions on Automatic Control,* Vol. 6, 1961, pp. 167-172.

[169]. D. Gabor, W. P. L. Wilby, and R. Woodcock, A universal nonlinear filter, predictor and simulator which optimizes itself by a learning process, *IRE Transactions on Automatic Control,* Vol. 6, 1961, pp. 422-438.

[170]. R. J. McGrath, V. Rajaraman, and V. C. Rideout, A parameter perturbation adaptive control system, *IRE Transactions on Automatic Control,* Vol. 6, 1961, pp. 154-161.

[171]. K. S. Narendra and L. E. McBride, Multiparameter self-optimizating systems using correlation techniques, *IEEE Transactions on Automatic Control,* Vol. 9, 1964, pp. 31-38.

[172]. J. B. Cruz Jr, System Sensitivity Analysis, *Hutchinson& Ross Inc,* Stroudsburg, PA, 1973.

[173]. H. P. Whitaker, J. Yamron, and A. Kezer, Design of model-reference adaptive control systems for aircraft, Instrumentation Laboratory, *MIT,* Cambridge, MA1958.

[174]. G. Goodwin and K. S. Sin, Adaptive Filtering Prediction and Control, *Prentice-Hall,* Englewood Cliffs, NJ, 1984.

[175]. B. D. O. Anderson, R R. Bitmead, C. R. Johnson, P. V. Kokotovic, R. L. Kosut, I. M. Y. Mareels, L. Praley, and B. D. Riedle, Stability of Adaptive Systems: Passivity and averaging analysis, *MIT Press,* Cambridge, MA, 1986.

[176]. D. Liu, A. J. McDaid, K. C. Aw, and S. Q. Xie, Position control of an Ionic Polymer Metal Composite actuated rotary joint using Iterative Feedback Tuning, *Mechatronics*, Vol. 21, 2011, pp. 315-328.

[177]. D. J. A. L. Leith, W. E., Survey of Gain-Scheduling Analysis and Design, *International Journal of Control*, Vol. 18, 2000, pp. 1001-1025.

[178]. W. J. Rugh and J. S. Shamma, Research on Gain Scheduling, *Automatica*, Vol. 36, 2000, pp. 1401–1425.

[179]. W. J. Rugh, Analytical Framework for Gain Scheduling, *IEEE Control System Magazine*, Vol. 11, 1991, pp. 79–84.

[180]. J. S. a. A. Shamma, M., Gain Scheduling: Potential Hazards and Possible Remedies, *IEEE Control System Magazine*, Vol. 12, 1992, pp. 101–107.

[181]. J. S. a. A. Shamma, M., Analysis of Gain Scheduled Control for Nonlinear Plants, *IEEE Trans. on Automatic Control*, Vol. 35, 1990, pp. 898–907.

[182]. R. A. Nichols, Reichert, R. T. and Rugh, W. J., Gain Scheduling for H1 Controllers: A Flight Control Example, *IEEE Trans. Control Systems Technology*, Vol. 1, 1993, pp. 69-79.

[183]. R. A. a. G. Hyde, K., The Application of Scheduled H1 Controllers to a VSTOL Aircraft, *IEEE Trans. on Automatic Control*, Vol. 38, 1993, pp. 1021–1039.

[184]. J. A. A. Shamma, M., Guaranteed Properties of Gain Scheduled Control of Linear Parameter-varying Plants, *Automatica*, Vol. 27, 1991, pp. 559–564.

[185]. A. J. McDaid, K. C. Aw, E. Haemmerle, and S. Q. Xie, Control of IPMC actuators for micro-fluidics with adaptive 'online' iterative feedback tuning, *IEEE/ASME Transactions on Mechatronics*, 2011.

[186]. K. C. Aw, W. Yu, A. J. McDaid, and S. Q. Xie, An IPMC Driven Micropump with Adaptive On-line Iterative Feedback Tuning, in *Proceedings of the 3rd International Conference on Smart Materials and Nanotechnology in Engineering*, Shenzhen, Guangdong, China 2011.

[187]. N. T. Nguyen, X. Huang, and T. K. Chuan, MEMS-micropumps: A review, *J. of Fluids Engineering, Trans. of the ASME*, Vol. 124, 2002, pp. 384-392.

[188]. V. Singhal, S. V. Garimella, and J. Y. Murthy, Low Reynolds number flow through nozzle-diffuser elements in valveless micropumps, *Sensors and Actuators, A, Physical*, Vol. 113, 2004, pp. 226-235.

[189]. J. G. Ziegler and N. B. Nichols, Optimum settings for automatic controllers, *Trans. Amer. Soc. Mech. Eng*, Vol. 64, 1942, pp. 759–768.

[190]. D. E. Seborg, T. F. Edgar, and D. A. Mellichamp, Process Dynamics and Control, *Wiley*, New York, 1989.

[191]. W. Yu, IPMC Micropump (Needs updating), Master's Thesis, Mechatronics Engineering, *The University of Auckland*, NZ, Auckland, 2011.

[192]. K. Inoue, T. Tanikawa, and T. Arai, Micro-manipulation system with a two-fingered micro-hand and its potential application in bioscience, *Journal of Biotechnology*, Vol. 133, 2008, pp. 219-224.

[193]. T. Yih and I. E. Talpasanu, Micro and Nano Manipulations for Biomedical Applications. Boston, *Artech House, Inc*, 2008.

[194]. B. Asiyanbola and W. Soboyejo, For the Surgeon: An Introduction to Nanotechnology, *Journal of Surgical Education*, Vol. 65, pp. 155-161.

[195]. L. Zhe and *et al.*, A micromanipulation system with dynamic force-feedback for automatic batch microinjection, *Journal of Micromechanics and Microengineering*, Vol. 17, 2007, p. 314.

[196]. M. Araki and H. Taguchi, Two-Degree-of-Freedom PID Controllers, *International Journal of Control, Automation and Systems*, Vol. 1, 2003, pp. 401-411.

[197]. D. H. Kim, B. Kim, and H. Kang, Development of a piezoelectric polymer-based sensorized microgripper for microassembly and micromanipulation, *Microsystem Technologies*, Vol. 10, 2004, pp. 275-280.

[198]. S. Yantao, X. Ning, and L. Wen Jung, Force-guided assembly of micro mirrors, in *Proceedings of the IEEE/RSJ International Conference on Intelligent Robots and Systems (IROS'03)*, Vol. 3, 2003, pp. 2149-2154.

[199]. C. Ho-Yin and W. J. Li, A thermally actuated polymer micro robotic gripper for manipulation of biological cells, in *Proceedings of the IEEE International Conference on Robotics and Automation (ICRA'03)*, Vol. 1, 2003, pp. 288-293.

[200]. A. Ashkin, Optical trapping and manipulation of neutral particles using lasers, In *Proceedings of the National Academy of Sciences*, Vol. 94, 1997, pp. 4853-4860.

[201]. F. Arai, T. Sakami, K. Yoshikawa, H. Maruyama, and T. Fukuda, Synchronized laser micromanipulation of microtools for assembly of microbeads and indirect manipulation of microbe, in *Proceedings of the IEEE/RSJ International Conference on Intelligent Robots and Systems (IROS' 2003)*, Vol. 3, 2003, pp. 2121-2126.

[202]. S. Masuda, M. Washizu, and I. Kawabata, Movement of blood cells in liquid by nonuniform traveling field, *IEEE Transactions on Industry Applications*, Vol. 24, 1988, pp. 217-222.

[203]. G. Fuhr, R. Hagedorn, T. Muller, B. Wagner, and W. Benecke, Linear motion of dielectric particles and living cells in microfabricated structures induced by traveling electric fields, in *Proceedings of the IEEE Conference on Micro Electro Mechanical Systems (MEMS '91): An Investigation of Micro Structures, Sensors, Actuators, Machines and Robots*, 1991, pp. 259-264.

[204]. M. Takeuchi, M. Nakajima, M. Kojima, and T. Fukuda, Soft handling probe using thermal gel for single cells, in *Proceedings of the International Symposium on Micro-NanoMechatronics and Human Science (MHS)*, 2010, pp. 311-316.

[205]. P. Jungyul, J. Seng-Hwan, K. Young-Ho, K. Byungkyu, L. Seung-Ki, J. Byungkwon, and L. Kyo-Il, An integrated bio cell processor for single embryo cell manipulation, in *Proceedings of the IEEE/RSJ International Conference on Intelligent Robots and Systems (IROS'04)*, 2004, Vol. 1, pp. 242-247.

[206]. J. H. Shim, H. S. Cho, and S. Kim, An actively compliable probing system, *IEEE Control Systems Magazine*, Vol. 17, 1997, pp. 14-21.

[207]. U. Deole and R. Lumina, Measuring the Load-Carrying Capability of IPMC Microgripper Fingers, in *Proceedings of the IEEE 32nd Annual Conference on Industrial Electronics (IECON'06)*, 2006, pp. 2933-2938.

[208]. K. Yun and W. J. Kim, System identification and microposition control of ionic polymer metal composite for three-finger gripper manipulation, in *Proceedings of the Institution of Mechanical Engineers. Part I: Journal of Systems and Control Engineering*, Vol. 220, 2006, pp. 539-551.

[209]. Z. Chen, Y. Shen, J. Malinak, N. Xi, and X. Tan, Hybrid IPMC/PVDF structure for simultaneous actuation and sensing, in *Proceedings of the SPIE - The International Society for Optical Engineering,* 2006.

[210]. K. Deok-Ho, Y. Seok, and K. Byungkyu, Mechanical force response of single living cells using a microrobotic system, in *Proceedings of the IEEE International Conference on Robotics and Automation (ICRA'04),* Vol. 5, 2004, pp. 5013-5018.

[211]. I. M. Horowitz, Synthesis of Feedback Systems, *Academic Press,* 1963.

[212]. Intuitive Surgical Announces Fourth Quarter Earnings, http://www.intuitivesurgical.com

[213]. D. R. Yates, M. Rouprêt, M. O. Bitker and C. Vaessen, To infinity and beyond: The robotic toy story, *European Urology,* Vol. 60, 2011, pp. 263-265.

[214]. U. Seibold, and G. Hirzinger, A 6-Axis Force/Torque Sensor Design for Haptic Feedback in Minimally Invasive Robotic Surgery, in *Proceedings of the 2nd VDE World Microtechnologies Congress,* Munich, 2003, pp. 14-15.

[215]. McDaid, A. J., Xie, S. Q., and Aw, K. C., A compliant surgical robotic instrument with integrated IPMC sensing and actuation, *International Journal of Smart and Nano Materials,* Vol. 3, pp. 188-203, 2012.

[216]. K. K. Leang, S. Yingfeng, S. Song, and K. J. Kim, Integrated Sensing for IPMC Actuators Using Strain Gages for Underwater Applications, *IEEE/ASME Transactions on Mechatronics,* Vol. 17, 2012, pp. 345-355.

[217]. N. Bhat and W. J. Kim, Precision force and position control of an ionic polymer metal composite, in *Proceedings of the Institution of Mechanical Engineers. Part I: Journal of Systems and Control Engineering,* Vol. 218, 2004, pp. 421-432.

[218]. R. C. Richardson, M. C. Levesley, M. D. Brown, J. A. Hawkes, K. Watterson and P. G. Walker, Control of ionic polymer metal composites, *IEEE/ASME Transactions on Mechatronics,* Vol. 8, 2003, pp. 245-253.

[219]. A. J. McDaid, E. Haemmerle, S. Q. Xie and K. C. Aw, Design, Analysis, and Control of a Novel Safe Cell Micromanipulation System With IPMC Actuators, *J. Mech. Des.,* Vol. 135, 2013, pp. 061003-1 - 061003-10.

[220]. Andrew J. McDaid, Kean C. Aw, Sheng Q. Xie and Enrico Haemmerle, Robust control of a cell micro-manipulation system with IPMC actuators, *Int. J. Intelligent Systems Technologies and Applications* (in press).

[221]. Andrew J. McDaid, Kean C. Aw, Sheng Q. Xie and Enrico Haemmerle, Optimal force control of an IPMC actuated micromanipulator for safe cell handling, *In Proceedings of the SPIE - The International Society for Optical Engineering*, 8409, 2012, art. No. 84090J.

[222]. L. Fu, A. J. McDaid, K. C. Aw, A force compliant surgical robotic tool with IPMC actuator and integrated sensing, in *Proceedings of the 4th International Conference on Smart Materials and Nanotechnology in Engineering (SMN)*, 10 – 12 July, Gold Coast, Australia, 2013.

[223]. L. Fu, A. J. McDaid, K. C. Aw, Control of an IPMC Actuated Robotic Surgical Tool with Embedded Interaction Sensing, Control of an IPMC Actuated Robotic Surgical Tool with Embedded Interaction Sensing, in *Proceedings of the IEEE/ASME International Conference on Advanced Intelligent Mechatronics (AIM)*, Wollongong, Australia, 2013.

Index

www.ingramcontent.com/pod-product-compliance
Lightning Source LLC
Chambersburg PA
CBHW050457190326
41458CB00005B/1319